Counting and Probability

Third Edition

STEVEN ROMAN

Emeritus Professor
Department of Mathematics
California State University, Fullerton

Ordering: 949-854-5667
7/9/2010

Counting and Probability
ISBN 1-878015-18-4

Preface

This is one in a series of mathematics modules designed for the general college level student, whose background may include only intermediate level algebra. It can be used, together with other modules, or as a supplement to another text, in courses such as college algebra, liberal arts mathematics, finite mathematics, or discrete mathematics.

The subject of this module is elementary counting and probability. In Chapter 1 we discuss the Fundamental Counting Principle (also known as the Multiplication Rule.) Then we discuss permutations and combinations.

Chapter 2 is devoted to the fundamentals of elementary probability. We discuss sample spaces, the concept of probability and how to compute the probability of an event. In Section 2.2, we discuss the principle of inclusion-exclusion for two events, the notion of independent events, and then apply this to computing the number of successes in binomial experiments. In Section 2.3, we discuss conditional probability and expected values.

Chapter 3 is devoted to applications of probability. In Section 3.1, we apply our knowledge of probability to analyzing the simplest aspects of draw poker, as well as to the California State Lottery. In Section 3.2, we discuss how probability can be applied to the study of elementary heredity. The two sections of Chapter 3 are independent of one another, as are the subsections within Section 3.1. (Thus, one could cover poker, but not the lottery, or vice-versa.)

Students are strongly urged to obtain a calculator for use with this module. We feel that it is important to include examples and exercises of a realistic nature, and this requires a fairly large amount of arithmetical computation at times.

Answers to the odd numbered exercises are given in the back of the module, and answers to all of the exercises are available to the instructor upon request.

Preface to the Series

This series is designed to provide textbook material for a course in contemporary mathematics for college level students. For one reason or another, large publishers have not responded to the educational concerns of instructors. Through the use of desktop publishing techniques, this series of modules can provide the needed flexibility to adapt to the differing concerns of instructors and motivational needs of students. In particular, by using these modules, instructors can now select topics on a class-by-class basis. We hope that this series will provide both students and instructors with an adaptable learning tool to help increase enthusiasm for mathematics in the classroom.

Acknowledgments

I would like to express my indebtedness to Professor John G. Pierce for making a number of constructive suggestions for improvements in this module. Also, I am indebted to Ms. Donna Dolan and Ms. Joan Sholars for carefully proofreading the module and working all of the exercises.

Changes for the Third Edition

The changes for the third edition of this module are strictly cosmetic. We are now preparing our modules in EXP for Windows, Version 5.0. As a result, the page location of items may have changed, but there is no content change from the *Alternate* Second Edition.

Contents

Introduction

This module is an introduction to the subjects of counting and probability. Now you might be thinking that you learned how to count in grade school, and that was the end of that. However, it happens that counting can be a lot more involved than what you learned in grade school. There are many very important questions that require sophisticated counting techniques and, in fact, many important counting problems have never even been solved!

We won't be getting involved with unsolved problems in this module, but we will spend some time learning how to count such things as the number of possible automobile license plates in your state, the number of ways to arrange books on a bookshelf, the number of possible hands in your favorite card game, and on a more serious note, the number of possible banana splits that can be made at your local ice cream parlor. Even if these questions are not of earthshaking importance, they will introduce you to the art of counting, and if you don't take it too seriously, you might have some fun as well.

After introducing you to counting techniques in Chapter 1, we will discuss a very important application of these techniques, namely, to the theory of probability. We will begin with some of the special terminology of probability theory, and then discuss some commonly occurring misconceptions concerning probability. For instance, in the California State Lottery (and this is typical of other lotteries as well), you pick 6 numbers from among the numbers 1 through 53. If you guess all 6 numbers correctly, you win millions. Would you ever consider picking the numbers 1, 2, 3, 4, 5 and 6? Most people probably would not.

However, as we will learn, the laws of probability tell us that the numbers 1, 2, 3, 4, 5 and 6 are just as likely to occur as any other set of 6 numbers. You shouldn't consider this as surprising, but rather as a strong indication of how unlikely it is to win the California State Lottery!

As another example of a common misconception concerning the laws of probability, imagine that you have tossed a balanced coin 99 times, and that remarkably, it has come up heads each time. Would you be willing to bet that it will come up heads on the next toss? Many people would bet only on tails, figuring that tails are "way overdue."

However, the laws of probability tell us that, assuming the coin is evenly balanced, it is just as likely for the 100–th toss to come up heads as tails.

After discussing the main points of elementary probability, we will turn to some applications. Here, we analyze a simple version of draw poker (explaining all of the necessary rules of the game at that time) and then discuss the

California State Lottery, to give you an idea what to expect when you play. Then we discuss how probability theory plays a key role in elementary genetics.

We strongly urge that you obtain a good calculator, that can display at least 10 digits. Many of our examples and exercises use real–life numbers, and so there will on occasion be a fairly large amount of arithmetic computation. Therefore, a calculator will prove indispensable. (A calculator with a so–called *factorial key*, which should be marked [n!], or something similar, is preferable, but not essential.)

Chapter 1
Elementary Counting

1.1 The Fundamental Counting Principal

We begin our study of counting techniques with a very simple, yet very powerful rule, known as the *fundamental counting principle.* In fact, most of what we will do in this module is based on this simple rule.

Many counting problems amount to counting the number of ways to perform a certain finite sequence of tasks, where each task can be performed in several different ways. Let us consider a simple example.

Example 1

The State of California used to issue license plates consisting of 3 digits, followed by 3 letters. In order to count the total number of possible license plates, we observe that forming a license plate of this type can be thought of as performing a sequence of 6 tasks. The first task is to choose a digit for the first position in the license plate, the second task is to choose a digit for the second position, and the third task is to choose a digit for the third position. The fourth task is to choose a letter for the fourth position on the license plate, and so on.

Now, loosely speaking, the fundamental counting principle says that the number of ways to perform a sequence of tasks, such as these, is the *product* of the number of ways to perform each task. Since there are 10 possible digits, there are 10 ways to perform each of the first three tasks, and since there are 26 letters, there are 26 ways to perform each of the last three tasks. Hence, the total number of ways to perform all six tasks, that is, the total number of license plates, is

$$10 \cdot 10 \cdot 10 \cdot 26 \cdot 26 \cdot 26 = 17,576,000$$

It is interesting to note that currently, there are over $20,000,000$ automobiles registered in California. This is why California now issues license plates that consist of 1 digit, followed by 3 letters, followed by 3 digits. (How many license plates of this type are there?) ★

Before we can formally state the fundamental counting principle, we must discuss one important point. This principle applies only under the assumption that the number of ways to perform any of the tasks in the sequence does not depend on how the previous tasks were performed. In Example 1, for instance, the number of ways to choose a digit for the second position does not depend on which digit was chosen for the first position. In a sense, this is an assumption about the independence of the tasks in the sequence. If a sequence of tasks does not have this type of independence, then we simply cannot apply the fundamental counting principle.

With this in mind, let us now state the fundamental counting principle.

The Fundamental Counting Principle

Let $T_1, T_2, \ldots, T_k$ be a sequence of tasks, with the property that the number of ways to perform any task in the sequence does not depend on how the previous tasks in the sequence were performed. Then, if there are n_i ways to perform the i-th task T_i, for all $i = 1, 2, \ldots, k$ the number of ways to perform the entire sequence of tasks is the product $n_1 n_2 \cdots n_k$. ★

You may be wondering why the fundamental counting principle tells us to *multiply* the numbers $n_1, n_2, \ldots, n_k$. To see why, let us look at a simple example.

Example 2

Suppose we want to count the number of ways to choose a letter, followed by a digit. If we choose the letter a, then we can choose the digit in 10 different ways; if we choose the letter b, then we can again choose the digit in 10 different ways, and so on. Hence, we get a total of

$$\underbrace{10 + 10 + \cdots + 10}_{26 \text{ terms}} = 26 \cdot 10 = 260$$

different ways to perform these tasks. This shows that the number of ways to perform these two tasks is the *product* of 26 and 10. This same reasoning applies, of course, to the case of more than two tasks.

On the other hand, the number of ways to choose *either* a letter *or* a number, but not both, is the sum $26 + 10 = 36$. ★

Let us consider some additional examples of the fundamental counting principle.

Example 3

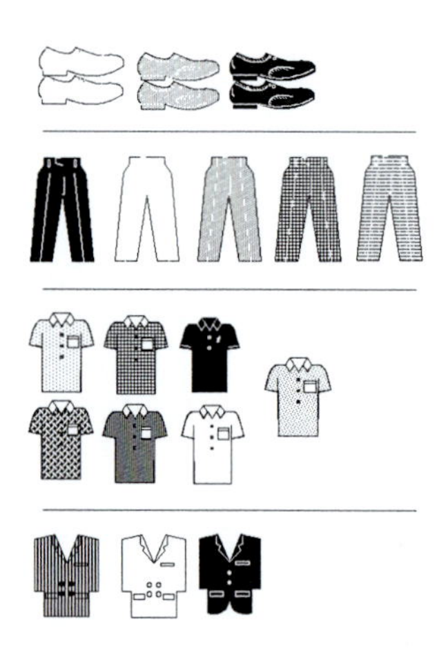

Suppose that you own 3 pairs of shoes, 5 pairs of pants, 7 shirts and 3 jackets. How many different outfits can you make out of these articles of clothing? (An outfit consists of 1 pair of shoes, 1 pair of pants, 1 shirt and 1 jacket.)

Solution To answer this question, we think of putting together an outfit as performing a sequence of 4 tasks. The first task is to pick a pair of shoes, the second task is to pick a pair of pants, the third task is to pick a shirt, and the fourth task is to pick a jacket. Since there are 3 ways to perform the first task, 5 ways to perform the second task, 7 ways to perform the third task, and 3 ways to perform the fourth task, the fundamental counting principle tells us that the number of ways to perform all four tasks, that is, the number of possible outfits, is

$$3 \cdot 5 \cdot 7 \cdot 3 = 315$$

Incidentally, the order in which we perform the 4 tasks does not matter, since we get the same outfit regardless of the order in which the shoes, pants, shirt and jacket are chosen. ★

Now let us do a serious example.

Example 4

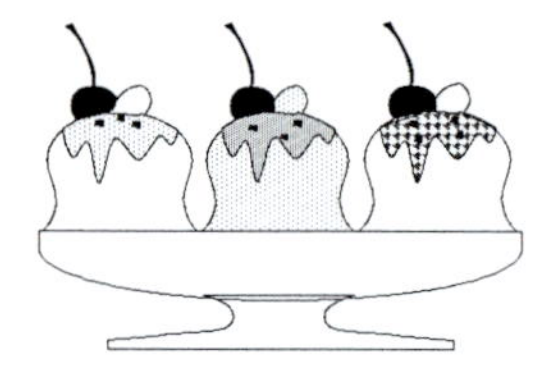

A banana split consists of 3 scoops of ice cream. Furthermore, each scoop may have one flavor of topping, and it may or may not have nuts and it may or may not have a cherry on top. If a certain ice cream parlor has 31 flavors of ice cream to choose from, and 5 different syrups, how many different banana splits can be made? We do distinguish order (from left to right) of the scoops in the dish. Thus, vanilla-vanilla-chocolate is different from chocolate-vanilla-vanilla.

Solution In this case, we think of making a banana split as performing a sequence of tasks, as follows

TASK 1: Choose a flavor of ice cream for the first scoop – 31 ways
TASK 2: Choose a flavor of ice cream for the second scoop – 31 ways
TASK 3: Choose a flavor of ice cream for the third scoop – 31 ways
TASK 4: Choose a topping for the first scoop – 5 ways
TASK 5: Choose a topping for the second scoop – 5 ways
TASK 6: Choose a topping for the third scoop – 5 ways
TASK 7: Choose either nuts or no nuts for the first scoop – 2 ways
TASK 8: Choose either nuts or no nuts for the second scoop – 2 ways
TASK 9: Choose either nuts or no nuts for the third scoop – 2 ways
TASK 10: Choose a cherry or no cherry for the first scoop – 2 ways
TASK 11: Choose a cherry or no cherry for the second scoop – 2 ways
TASK 12: Choose a cherry or no cherry for the third scoop – 2 ways

Thus, according to the fundamental counting principle (and with the help of a calculator), we see that there are a total of

$$31^3 \cdot 5^3 \cdot 2^6 = 238,328,000$$

possible banana splits. (Happy eating!) ★

Example 5

We have seen in Example 1 that there are $17,576,000$ license plates consisting of 3 digits followed by 3 letters. How many license plates of this type are there with no repeated digits or letters? (For instance, the license plate 456 ANA has a repeated letter, and thus should not be counted.)

Solution In this case, the task of choosing a digit for the first position can be performed in 10 ways. Regardless of the outcome of this task, there are only 9 remaining digits to choose from for the second position. Once the second position has been filled, there are only 8 digits to choose from for the third position. Similarly, there are 26 ways to choose a letter for the fourth position, 25 ways to choose a letter for the fifth position, and 24 ways to choose a letter for the last position. Hence, the total number of license plates with no repeated digits or letters is

$$10 \cdot 9 \cdot 8 \cdot 26 \cdot 25 \cdot 24 = 11,232,000$$

It is interesting to notice that the percentage of license plates with no repeated digits or letters is

$$\frac{11232000}{17576000} \approx 0.64$$

and so approximately 36%, or over one-third, of all license plates have a repeated digit or letter. ★

Example 6

Suppose you draw 3 cards from an ordinary deck of 52 cards. If the order in which the cards are drawn matters, how many different possibilities are there

a) if the cards are replaced after they are drawn, so that the same card can be drawn more than once?

b) if the cards are not replaced after each is drawn, so that the same card cannot be drawn more than once?

Solutions To answer the first question, observe that, since the cards are replaced after each one is drawn, there are always 52 possibilities for each draw. Hence, there are 52 ways to perform each of the 3 tasks of drawing a card from the deck, and so the answer to the first question is

$$52 \cdot 52 \cdot 52 = 140,608$$

As to the second question, there are 52 ways to draw the first card, but since that card is not replaced, there are only 51 ways to draw the second card, and similarly, there are only 50 ways to draw the third card. Hence, the total number of ways to draw 3 cards under these conditions is

$$52 \cdot 51 \cdot 50 = 132,600$$ ★

The fundamental counting principle is really a very simple idea – the trick is in recognizing when it applies, that is, in recognizing whether or not a given problem can be thought of as a problem involving a sequence of tasks. As we will see, a great many problems can be expressed in this way, and so this rule does apply in a surprisingly large number of problems.

EXERCISES

Define or discuss the following terms.

a) Fundamental counting principle

1. Count the number of license plates possible in your state, or in a neighboring state.
2. A certain state has driver's license numbers that consist of 1 letter followed by 6 digits. How many different driver's license numbers are possible in that state?
3. Suppose you own 6 jackets, 4 pairs of pants, 3 sweaters and 10 shirts. How many different outfits do you have?
4. Referring to Example 4, how many banana splits can be made up if the 3 scoops of ice cream must all be a different flavor?
5. Suppose that your schedule for next semester must consist of one natural science class, one liberal arts class, one humanities class, and one physical science class. How many ways can you make up your schedule if you can choose from 3 natural science classes, 4 liberal arts classes, 6 humanities classes and 3 physical science classes?
6. Suppose that there are 3 automated teller machines and 4 gas stations in your neighborhood. One morning, you need to go first to an automated teller machine, and then to a gas station. In how many ways can you do this?
7. Your English professor has required that you buy any one of 5 books on a certain list, and your history teacher has required that you buy any two of 3 books on a certain list. In how many ways can you buy these books? (*Question*: Does order matter here?)
8. How many ways are there to select one man, one woman and one child from among 5 men, 7 women and 3 children?
9. A certain candy store sells 4 different kinds of chocolate bars. Each bar is available in 3 sizes, and also with or without nuts. How many different types of chocolate bars are available?
10. How many different telephone numbers can be made from the digits $0, 1, \ldots, 9$ if the first digit cannot be 0?
11. How many different telephone numbers are possible if we include a three digit area code as part of each number, where neither the first digit of the area code nor the first digit of the number itself can be 0?
12. A student is taking a 10 question true-false test, but doesn't know any of the answers. How many different possibilities are there for filling out the test?
13. A student is taking a multiple choice test consisting of 10 questions, each of which has 4 alternatives. In how many ways can the student fill out the test?
14. Suppose you decide to open a shoe store. You realize that you need to stock at least 25 different varieties of shoes, each in 10 different lengths and each length in 3 different widths. Also, each style and size of shoe must be stocked in 3 different colors. How many pairs of shoes must you stock, assuming no duplicates?
15. A single die is rolled and a single coin is flipped. How many different outcomes are possible?
16. Two dice (one red and one blue) are rolled and three coins (a penny, a nickel and a dime) are flipped. How many outcomes are possible?
17. Four tarot cards are drawn from a deck of 78 cards. How many outcomes are possible if the order in which the cards are drawn matters, and if

a) each card is replaced before the next one is drawn?
b) the cards are not replaced after they are drawn?

18. If you flip a coin 4 times and record the sequence of heads and tails how many possibilities are there for this sequence?
19. Suppose you are given 3 boxes. Box 1 contains 5 objects, box 2 contains 4 objects, and box 3 contains 2 objects.
    a) How many ways are there to choose exactly one object?
    b) How many ways are there to choose exactly one object from each box?
20. Suppose you are given a box that contains 10 balls, numbered 1 through 10. You select 3 of these balls, and record their numbers, obtaining in this way a sequence of 3 numbers.
    a) How many such sequences of 3 numbers are there if you replace each ball as soon as you have recorded its number, thus allowing for the possibility of choosing the same ball more than once?
    b) How many such sequences of 3 numbers are there if you do not replace a ball once it is chosen?
21. A frozen yogurt store in your neighborhood sells 5 flavors of frozen yogurt, 6 different syrups and 8 different dry toppings.
    a) How many sundaes are possible using one flavor of yogurt, one type of syrup and one dry topping?
    b) How many sundaes are possible using one flavor of yogurt, one syrup and two different dry toppings? (Be careful here.)
22. How many 4-digit numbers are there, whose digits come from $0, 1, 2, 3, 4$ and 5?
23. How many 4-digit numbers are there, of the type described in Exercise 22, that are also even?
24. How many 4-digit numbers are there, of the type described in Exercise 22, that are also a multiple of 5?

1.2 Permutations

Let us consider the following counting problem. Suppose that you are the manager of a small art gallery. To prepare for an exhibit, you must hang 5 paintings by the same artist in a row on one wall, as shown in Figure 1.

Figure 1

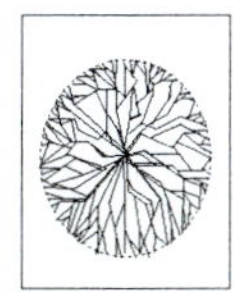
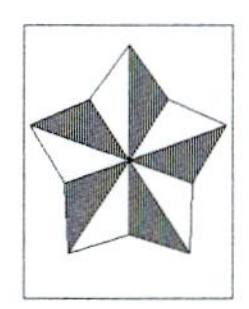
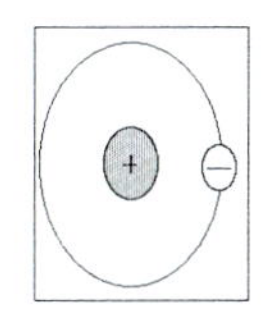
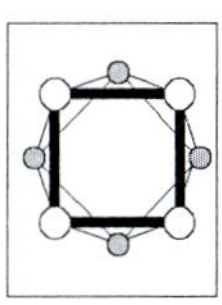

$$p_1, p_2, p_3, p_4, p_5$$

The artist is very temperamental and cares in which order you hang the paintings, but he neglected to tell you the order he prefers. The question is "How many orders are possible?"

If we denote the 5 paintings by the symbols p_1, p_2, p_3, p_4 and p_5, then each possible order can be represented by an arrangement of these symbols. For example, the arrangement shown in Figure 1 is

$$p_1 p_2 p_3 p_4 p_5$$

Figure 2 shows another possible arrangement, namely

$$p_3 p_5 p_1 p_2 p_4$$

Figure 2

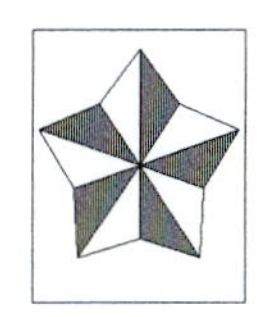
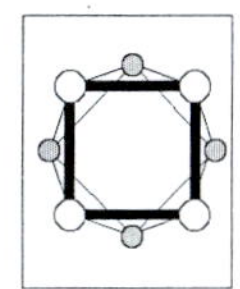

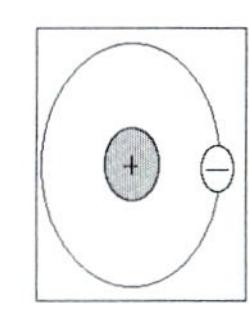

$$p_3, p_5, p_1, p_2, p_4$$

Ordered arrangements of objects occur very frequently, and they have a special name.

Definition

An ordered arrangement of a set of objects is called a **permutation** of the objects. If there are n objects in the permutation, we say that the permutation has **size** n. ★

Jacques Bernoulli
1654-1705
The first to use the term permutation in its current meaning.

Thus, in this case, each arrangement of the paintings is a permutation of size 5, and we are interested in knowing how many permutations there are of size 5, or indeed, of any size.

Before deriving a formula for the number of permutations of size n, let us consider a few more examples of permutations.

Example 1

There are exactly 2 permutations of the two letters a and b , namely

$$ab \text{ and } ba$$

There are 6 permutations of the three letters a, b and c,

Beginning with a:	abc	acb
Beginning with b:	bac	bca
Beginning with c:	cab	cba

There are 24 permutations of the four letters a, b, c and d. These are

Beginning with a:	$abcd$	$abdc$	$acbd$	$acdb$	$adbc$	$adcb$
Beginning with b:	$bacd$	$badc$	$bcad$	$bcda$	$bdac$	$bdca$
Beginning with c:	$cabd$	$cadb$	$cbad$	$cbda$	$cdab$	$cdba$
Beginning with d:	$dabc$	$dacb$	$dbac$	$dbca$	$dcab$	$dcba$ ★

As you can see, there are a rather large number of permutations of size n, even for relatively small values of n. This makes it especially important that we have a formula for this number. In order to state this formula, we need a bit of notation. For n a positive integer, we use the notation $n!$ to denote the product of the first n positive integers. The expression $n!$ is read "n factorial." Some examples should make this clear.

$1! = 1$

$$2! = 1 \times 2 = 2$$
$$3! = 1 \times 2 \times 3 = 6$$
$$4! = 1 \times 2 \times 3 \times 4 = 24$$
$$5! = 1 \times 2 \times 3 \times 4 \times 5 = 120$$

It is customary to set $0!$ equal to 1. For reference, Table 1 lists the values of the first few factorials.

Table 1	
$0! = 1$	$6! = 720$
$1! = 1$	$7! = 5040$
$2! = 2$	$8! = 40,320$
$3! = 6$	$9! = 362,880$
$4! = 24$	$10! = 3,628,800$
$5! = 120$	$11! = 39,916,800$

Now we can use the fundamental counting principle to derive a formula for the number of permutations of size n.

Theorem 1

There are $n!$ permutations of size n.

Proof. Forming a permutation of n objects can be thought of as performing a sequence of n tasks. The first task is to choose the first object for the permutation, and since there are n objects to choose from, this can be done in n different ways. The second task is to choose the second object for the permutation, and since *regardless of the outcome of the first task*, there are always $n-1$ objects remaining, the second task can be performed in $n-1$ ways. The third task is to choose the third object for the permutation, and this can be done in $n-2$ ways, regardless of the outcome of the first two tasks. This continues until we reach the n-th task, which is to choose the n-th object for the permutation, and this can be done in only one way, since there is only one object left at this point.

Thus, according to the fundamental counting principle, there are

$$n(n-1)\cdots 1 = n!$$

different ways to perform the entire sequence of n tasks, that is, there are $n!$ permutations of n objects. This completes the proof. ★

Using Theorem 1, we can easily solve the problem of how many ways there are to arrange 5 paintings in a row on a wall. For, according to this theorem, there are $5! = 120$ permutations of size 5, and so there are 120 ways to arrange the paintings. Here is another example.

Example 2

Suppose that you want to arrange 2 math books, 3 English books, and 2 history books on a bookshelf.

a) In how many ways can this be done?
b) In how many ways can this be done if the math books must come first, then the English books, and finally the history books?
c) In how many ways can this be done if all of the books of the same subject must be kept together?

Solutions

a) To answer part a), we observe that each arrangement of the 7 books on the bookshelf is a permutation of the books. Hence, according to Theorem 1, and Table 1, there are

$$7! = 5040$$

possible arrangements of the books.

b) Since there are 2! different ways to arrange the math books, 3! different ways to arrange the English books, and 2! different ways to arrange the history books, the fundamental counting principle tells us that there are

$$2!3!2! = 2 \cdot 6 \cdot 2 = 24$$

different ways to arrange the books on the bookshelf, under these conditions.

c) In this case, placing the books on the bookshelf can be thought of as performing two tasks. The first task is to choose the order for the three *subjects* – math, English and history. This task can be performed in $3! = 6$ different ways. Once this task has been performed, we can use the results of part b) to tell us how many ways there are to place the actual books, which in this case is 24. Hence, the answer to part c) is $6 \cdot 24 = 144$. ★

Example 3

a) How many batting orders are possible, among 9 baseball players? (A batting order is simply an arrangement of the players, giving each player a specific turn at bat.)
b) How many batting orders are possible with the 9 players $p_1, p_2, \ldots, p_9$, if player p_2 cannot immediately follow player p_1 in the batting order?

Solutions

a) Since each batting order corresponds to a permutation of the 9 players, there are

$$9! = 362,880$$

possible batting orders.

b) This is an example of a type of problem where it is much easier to count what we do not want, and subtract that from the total. In this case, there are a total of 9! batting orders, and so if we subtract from 9! the number of batting orders in which player p_2 *does* follow player p_1, then we will have the answer to our question.

In order to compute the number of batting orders in which player p_2 does follow player p_1, we reason as follows. As long as player p_2 must follow player p_1, we can think of these two players as "tied together" into one "player." In effect then, there are only 8 players, and so there are 8! batting orders in which p_2 follows p_1. Thus, there are

$$9! - 8! = 362,880 - 40,320 = 322,560$$

batting orders in which player p_2 does not follow player p_1. ★

Permutations of Size k, Taken from n Objects

It often happens that we have a set of n objects, but that we want to form permutations using only k of the objects at a time, where $k < n$.

Example 4

The permutations of size 2, taken from the set $\{a, b, c, d\}$ of 4 letters, are

$ab \quad ba \quad ac$
$ca \quad ad \quad da$
$bc \quad cb \quad bd$
$db \quad cd \quad dc$

To list the permutations of size 3, taken from the set $\{a, b, c, d\}$, we group them according to which letters they involve

permutations involving	abc	acb	bac
the letters a , b and c	bca	cab	cba
permutations involving	abd	adb	bad

the letters a , b and d	bda	dab	dba
permutations involving	acd	adc	cad
the letters a , c and d	cda	dac	dca
permutations involving	bcd	bdc	cbd
the letters b , c and d	cdb	dbc	dcb ★

Using the fundamental counting principle, we can obtain a formula for the number of permutations of size k, taken from a set of n objects. Let us denote this number by $P(n, k)$.

Theorem 2

The number $P(n, k)$ of permutations of size k, taken from a set of n objects, is

$$P(n, k) = n(n-1)\cdots(n-k+1) \tag{1}$$

Proof. Forming a permutation of size k, from a set of n objects, can be thought of as performing a series of k tasks. The first task is to choose the first object for the permutation, and there are n ways to do this. The second task is to choose the second object for the permutation, and there are $n-1$ ways to do this, and so on. The k−th task is to choose the k−th object for the permutation. Now, when it comes time to perform the k−th task, we have already performed $k-1$ tasks, and so there are $n-(k-1) = n-k+1$ objects remaining. Hence, there are precisely $n-k+1$ ways to perform the k−th task, and so, according to the fundamental counting principle, we have

$$P(n, k) = n(n-1)\cdots(n-k+1)$$ ★

Perhaps the easiest way to evaluate the right hand side of equation (1) is to *first* compute $n-k+1$. Then, according to Theorem 2, $P(n, k)$ is equal to the product of consecutive integers from n *down* to $n-k+1$. Let us illustrate.

Example 5

The number of permutations of size 3, taken from a set of 4 objects, is $P(4, 3)$. To evaluate this using Theorem 2, we note that $n = 4$ and $k = 3$, and so $n-k+1 = 4-3+1 = 2$. Hence,

$$P(4, 3) = 4 \cdot 3 \cdot 2 = 24$$

Notice that this is in agreement with the previous example.

The number of permutations of size 4, taken from a set of 8 objects, is $P(8, 4)$. To evaluate this number, we again compute

$$n-k+1 = 8-4+1 = 5$$

Hence,

$$P(8, 4) = 8 \cdot 7 \cdot 6 \cdot 5 = 1680$$ ★

Formula (1) can also be written in terms of factorials by observing that

$$n(n-1)\cdots(n-k+1) = \frac{n(n-1)\cdots(n-k+1)(n-k)\cdots 1}{(n-k)\cdots 1} = \frac{n!}{(n-k)!}$$

and so

$$P(n,k) = \frac{n!}{(n-k)!} \tag{2}$$

Example 6

Suppose that the artist mentioned in the beginning of this section had painted 10 paintings, but only 5 will fit on the wall at one time. How many ways are there to hang any 5 of the artist's paintings?

Solution In this case, each arrangement of 5 of the 10 paintings corresponds to a permutation of size 5, taken from 10 objects. Hence, there are $P(10,5)$ possible arrangements of the paintings. To evaluate this number, we note that $n-k+1 = 10-5+1 = 6$, and so

$$P(10,5) = 10 \cdot 9 \cdot 8 \cdot 7 \cdot 6 = 30,240$$

Thus, there are 30, 240 ways to choose and arrange 5 of the 10 paintings. ★

EXERCISES

Define or discuss the following terms.

a) Ordered arrangement b) Permutation
c) Size of a permutation d) Factorial
e) Permutation of n objects taken k at a time

1. Compute the following values without looking at Table 1.
   a) $0!$ b) $1!$ c) $3!$
   d) $4!$ e) $5!$ f) $6!$
2. Compute the following values.
   a) $P(5,1)$ b) $P(5,2)$ c) $P(7,3)$
   d) $P(5,5)$ e) $P(6,0)$ f) $P(100,2)$
3. Use a calculator to compute the following values.
   a) $P(12,5)$ b) $P(20,8)$ c) $P(100,4)$
4. Compute and compare the following values without looking at Table 1.
   a) $3! - 1!$ and $(3-1)!$ b) $4! + 2!$ and $(4+2)!$
   c) $\frac{4!}{2!}$ and $\left(\frac{4}{2}\right)!$ d) $3! \cdot 2!$ and $(3 \cdot 2)!$
5. Simplify the expression
   a) $\frac{3!}{2!}$ b) $\frac{4!}{3!}$ c) $\frac{5!}{4!}$
   d) $\frac{n!}{(n-1)!}$ (*Hint*, observe the pattern in parts a–c.)
6. Are $(2n)!$ and $2n!$ the same for all positive integers n? Justify your answer.
7. Are $(n+m)!$ and $n! + m!$ the same for all positive integers n and m? Justify your answer.

8. Are $(nm)!$ and $n!m!$ the same for all positive integers n and m? Justify your answer.
9. Find the smallest value of n for which $n!$ is larger than the population of the world.
10. How many permutations are there of size 8?
11. How many permutations are there of elements in the set $\{1, 2\}$, using both elements? List them.
12. How many permutations are there of the elements in the set $\{1, x, y\}$, using all three elements? List them.
13. How many permutations are there of size 5, taken from a set of size 6?
14. How many permutations are there of size 5, taken from a set of size 8?
15. Determine the number of permutations of size 2, taken from the set $\{1, 2, 3, 4, 5\}$. List them.
16. In how many ways can 6 people line up at a supermarket checkout counter?
17. How many ways are there to arrange 10 people in a line? Suppose it took 1 minute to rearrange these 10 people in any order you desired. How many *years* would it take to try out all of the possibilities?
18. How many ways are there to arrange the first 25 exercises in this section?
19. How many ways are there to arrange the first 25 exercises in this section if the first five exercises must come first, but they can appear in any order?
20. How many permutations are there in the letters in your last name?
21. In a horse race involving 9 horses, how many ways can first, second and third places be decided?
22. How many batting orders are possible, with 9 players, if player p_3 must directly follow player p_2, who must directly follow player p_1?
23. How many batting orders are possible if player p_2 must follow player p_1, and player p_4 must follow player p_3?
24. A certain baseball team has 4 girls and 5 boys. How many batting orders are possible if the girls must bat first?
25. A certain baseball team has 4 boys and 4 girls. How many batting orders are possible if the sexes must alternate?
26. How many 5 letter words can be formed if no letter is allowed to be used more than once in any word? (A word is any combination of letters – it does not have to be meaningful.)
27. How many 5 letter words can be formed if no letter is allowed to be used more than once in any word, and if the letter q must always be followed by the letter u?
28. There are 24 letters in the Greek alphabet. If each fraternity and each sorority must choose a different three letter insignia, how many fraternities and sororities are possible?
29. How large can the population of a town be if no two people are allowed to have the same three letter initials?
30. How many ordered triples of letters are there, taken from the letters A, T, C and G
    a) if repeated letters are allowed?
    b) if repeated letters are not allowed?
31. Radio station call letters are made up of four letters, beginning with either a "K" or a "W."

a) How many call letters are possible if repetitions are allowed?
b) How many call letters are possible if repetitions are not allowed in any of the four letters?
c) How many call letters are possible if repetitions are not allowed in the last three letters?

32. Suppose you wish to arrange 2 math books, 5 chemistry books, and 4 history books on a single bookshelf.
    a) In how many ways can this be done?
    b) In how many ways can this be done if the math books must come first, then the chemistry books, and finally the history books?
    c) In how many ways can this be done if all of the books of the same subject must be kept together?
33. Suppose that you wish to arrange 5 math books, 5 chemistry books, and 5 English books on a single bookshelf. In how many ways can this be done if the subjects must alternate – first a math book, then a chemistry book, then an English book, and so on?
34. How many ways are there to hand out 7 free books among 15 students if no student is to receive more than 1 book?
35. How many ways are there to choose 5 people from a group of 10 people and place the 5 people in a line?
36. A combination for a combination lock consists of a sequence of 4 numbers between 0 and 59 (inclusive).
    a) How many possible combinations are there?
    b) How many possible combinations are there if no number can be repeated?
37. Suppose you have 8 red flowers and 8 white flowers, and assume that the flowers are indistinguishable except by their color.
    a) How many ways are there to arrange 8 of these flowers in a row in a flowerpot?
    b) How many ways are there to arrange 8 of these flowers in a row if the colors must alternate?
    c) How many ways are there to arrange 5 of the red flowers and 5 of the white flowers in a row if all of the red flowers must be kept together?
38. Given 7 beads of different colors, how many ways are there to make a necklace using these beads?

1.3 Combinations

In the previous section, we counted ordered arrangements of objects. In this section, we count unordered arrangements. You are probably already somewhat familiar with unordered arrangements, since an unordered arrangement is the same as a subset. Unordered arrangements are also referred to by another name.

Definition

An unordered arrangement, or subset, of size k, taken from a set of size n, is called a **combination** of size k. ★

We will use the terms subset and combination interchangeably. Also, since a combination is a subset, we will use set notation to denote combinations. For example, the set $\{1, 3, 5\}$ is a combination of size 3, taken from the set $S = \{1, 2, 3, 4, 5, 6\}$.

The number of combinations (or subsets) of size k, taken from a set of size n is denoted by $C(n, k)$. Let us consider some examples.

Example 1

For the sake of comparison, let us make a list of all of the combinations of size 3 and all of the permutations of size 3, taken from the set $\{a, b, c, d\}$.

	Combinations	*Permutations*		
Those involving	$\{a, b, c\}$	abc	acb	bac
a, b and c:		bca	cab	cba
Those involving	$\{a, b, d\}$	abd	adb	bad
a, b and d:		bda	dab	dba
Those involving	$\{a, c, d\}$	acd	adc	cad
a, c and d:		cda	dac	dca
Those involving	$\{b, c, d\}$	bcd	bdc	cbd
b, c and d:		cdb	dbc	dcb

Notice that each combination gives rise to an entire group of 6 permutations, namely, those permutations that involve the same letters as the given combination. This observation is the key to finding a formula for the numbers $C(n, k)$. ★

The next theorem describes a relationship between the numbers $C(n, k)$ and $P(n, k)$.

Theorem 1

The number $C(n, k)$ of combinations of size k, taken from a set of n objects, is given by the formula

$$C(n,k) = \frac{P(n,k)}{k!} \tag{1}$$

Proof. The proof amounts to nothing more than generalizing the observation that we made in the previous example. Imagine a set of n elements, and imagine that we have made a list of all combinations of size k, taken from that set. Each such combination gives rise to a whole group of permutations. In fact, for each combination of size k, there are $k!$ permutations – one permutation for each ordered arrangement of the elements in the combination.

This means that if we multiply the number of combinations of size k by $k!$, we should get the number of permutations of size k. In symbols,

$$C(n,k) \cdot k! = P(n,k)$$

Dividing both sides of this by $k!$ gives formula (1). ★

Using formula (1), together with formula (2) from the previous section, we get

$$C(n,k) = \frac{P(n,k)}{k!} = \frac{n!}{k!(n-k)!}$$

and so

$$C(n,k) = \frac{n!}{k!(n-k)!} \tag{2}$$

Let us try some examples of formulas (1) and (2).

Example 2

a) Find the number of subsets of size 3, taken from a set of 4 objects.
b) Find the number of subsets of size 5, taken from a set of 9 objects.

Solutions

a) According to formula (2), we have

$$C(4,3) = \frac{4!}{3!(4-3)!} = \frac{4!}{3! \cdot 1!} = \frac{24}{6 \cdot 1} = 4$$

which is in agreement with Example 1.

b) When the numbers n and k grow large, it is easier to evaluate $C(n,k)$ using formula (1). To do this, we need to first compute $P(9,5)$. Since $n-k+1 = 9-5+1 = 5$, we have

$$P(9,5) = 9 \cdot 8 \cdot 7 \cdot 6 \cdot 5$$

and so

$$C(9,5) = \frac{P(9,5)}{5!} = \frac{9 \cdot 8 \cdot 7 \cdot 6 \cdot 5}{5!} = 126$$

★

Example 3

In the game of draw poker, a hand consists of 5 cards. How many possible poker hands are there in an ordinary deck of 52 cards?

Solution The number of possible poker hands is $C(52, 5)$ which, according to formula (1), is

$$C(52,5) = \frac{P(52,5)}{5!} = \frac{52 \cdot 51 \cdot 50 \cdot 49 \cdot 48}{120} = 2,598,960$$

Hence, there are $2,598,960$ possible 5-card poker hands. ★

Even with a calculator, evaluating the numbers $C(n, k)$ for large values of n and k can sometimes be a bit tedious. However, the following formula can make the job much easier.

$$C(n,k) = C(n,n-k) \tag{3}$$

This formula can be justified by observing that it counts the same thing in two different ways. There are $C(n, k)$ ways to choose k objects from n objects. But we can also count this number by choosing the $n - k$ objects that we don't want, and discarding them. This can be done in $C(n, n - k)$ ways. Let us consider an example.

Example 4

Suppose that you are trying to arrange 100 photographs in an album that can only hold 95 photos. How many ways can you fill this album if the order in which you place the photographs in the album does not matter?

Solution Here we need to evaluate $C(100, 95)$. According to formula (3), we have

$$C(100,95) = C(100,100-95) = C(100,5)$$

To evaluate $C(100, 5)$, we use formula (1). Since $100 - 5 + 1 = 96$, a calculator gives

$$P(100,5) = 100 \cdot 99 \cdot 98 \cdot 97 \cdot 96 = 9,034,502,400$$

and so

$$C(100,95) = C(100,5) = \frac{P(100,5)}{5!} = \frac{9,034,502,400}{120} = 75,287,520$$

Thus, you have about 75 million ways to choose 95 photographs for your album! (How many ways are there to fill the album if you care about the order of the photographs?) ★

Example 5

A certain club has 4 male and 6 female members. How many ways are there to form a 5 person committee of club members consisting of 2 men and 3 women?

Solution There are $C(4,2)$ ways of choosing a subset of 2 of the 4 men to serve on the committee, and there are $C(6,3)$ ways to choose a subset of 3 of the 6 women to serve on the committee. Hence, according to the fundamental counting principle, there are

$$C(4,2)C(6,3) = 6 \cdot 20 = 120$$

ways to form the 5 person committee. ★

In the next chapter, there will be several times when we want to count the number of n-tuples of a certain type, so let us do an example of this kind now.

Example 6

a) How many 5-tuples are there, whose entries are either H or T?
b) How many of these 5-tuples have exactly two T's?

Solutions

a) Since there are 2 choices for each of the 5 positions in a 5-tuple, the fundamental counting principle tells us that there are

$$2 \cdot 2 \cdot 2 \cdot 2 \cdot 2 = 2^5 = 32$$

such 5-tuples.

b) To determine the number of 5-tuples containing exactly two T's, we can reason as follows. First we write an empty 5-tuple

$$(, , , ,)$$

Now, to construct a 5-tuple with exactly two T's, we simply pick 2 of the 5 empty positions in which to place the T's. Once the T's are in place, the remaining positions are filled with H's. Since there are $C(5,2)$ ways to pick 2 positions out of 5, there are $C(5,2) = 10$ 5-tuples with exactly two T's.★

The results of the previous example are so useful that we should generalize them in a theorem.

Theorem 2

a) There are 2^n n-tuples, whose entries are either H or T.
b) The number of such n-tuples that contain exactly k T's, is $C(n,k)$. ★

Of course, Theorem 2 applies to any situation where the entries of the n-tuples can be one of two values (not just H and T).

Blaise Pascal
(1623-1662)

The numbers $C(n,k)$ are among the most important numbers in mathematics, and their properties have been studied extensively. In 1653, the mathematician Blaise Pascal organized these numbers in the form of a triangle, which has come to be known as **Pascal's triangle**. (Actually, Pascal's triangle was known to the Chinese several centuries earlier, but Pascal was the first to make an extensive study of it.)

For convenience, we put Pascal's triangle into a table format. For instance, to find the value of $C(9,6)$, we look in the row labeled 9 ($n = 9$) and the column labeled 6 ($k = 6$), to get $C(9,6) = 84$.

Pascal's Triangle — Values of $C(n,k)$											
$n\backslash k$	0	1	2	3	4	5	6	7	8	9	10
0	1										
1	1	1									
2	1	2	1								
3	1	3	3	1							
4	1	4	6	4	1						
5	1	5	10	10	5	1					
6	1	6	15	20	15	6	1				
7	1	7	21	35	35	21	7	1			
8	1	8	28	56	70	56	28	8	1		
9	1	9	36	84	126	126	84	36	9	1	
10	1	10	45	120	210	252	210	120	45	10	1

Let us conclude by remarking that the numbers $C(n,k)$ are usually referred to as the **binomial coefficients**. This terminology comes from the fact that these numbers appear in the expansion of the expression $(x+y)^n$. For instance, as you know,

$$(x+y)^2 = x^2 + 2xy + y^2$$

Looking at the row of Pascal's triangle associated with $n=2$, we see that this formula can be written in the form

$$(x+y)^2 = C(2,0)x^2 + C(2,1)xy + C(2,2)y^2$$

This is no accident. We also have

$$(x+y)^3 = C(3,0)x^3 + C(3,1)x^2y + C(3,2)xy^2 + C(3,3)y^3$$

and

$$(x+y)^4 = C(4,0)x^4 + C(4,1)x^3y + C(4,2)x^2y^2 + C(4,3)xy^3 + C(4,4)y^4$$

To see why these coefficients are reasonable, let us look at the product

$$(x+y)^4 = (x+y)(x+y)(x+y)(x+y)$$

In expanding this product, we *choose* an x or a y from each factor. To get the coefficient of xy^3, for example, we choose 3 of the 4 factors to contribute a y, and the remaining factor contributes an x. Since there are $C(4,3)$ ways to choose 3 factors, we get $C(4,3)$ terms of the form xy^3. Collecting these terms together, we get $C(4,3)xy^3$.

A similar formula holds for $(x+y)^n$, for *all* positive integral powers n. These formulas are known collectively as the **binomial formula**. We mention these facts only so that you will be familiar with the term binomial coefficient. We will not have any direct use for the binomial formula in this module.

EXERCISES

Define or discuss the following terms.

a) Combination
b) Size of a combination
c) Pascal's triangle
d) Binomial coefficient
e) Binomial formula

In Exercises 1–18, evaluate the given expression, without looking at Pascal's triangle.

1. $C(2,1)$
2. $C(3,1)$
3. $C(3,2)$
4. $C(5,2)$
5. $C(5,4)$
6. $C(5,5)$
7. $C(6,1)$
8. $C(6,2)$
9. $C(6,3)$
10. $C(8,4)$
11. $C(8,5)$
12. $C(9,4)$
13. $C(9,7)$
14. $C(9,9)$
15. $C(10,3)$
16. $C(10,5)$
17. $C(10,6)$
18. $C(10,7)$

In Exercises 19 – 30, use a calculator if necessary to evaluate the given expression.

19. $C(15,5)$
20. $C(15,9)$
21. $C(15,14)$
22. $C(20,4)$
23. $C(20,7)$
24. $C(20,15)$
25. $C(75,6)$
26. $C(75,70)$
27. $C(75,74)$
28. $C(100,99)$
29. $C(100,98)$
30. $C(100,97)$
31. Verify that $C(n,0) = 1$ for all positive integers n.
32. Verify that $C(n,1) = n$ for all positive integers n.
33. Verify that $C(n,n) = 1$ for all positive integers n.
34. How many subsets are there of size 2, taken from the set $\{a,b,c,d\}$? List them.
35. How many subsets of size 3 are there, taken from the set $\{1,2,3,4,5\}$? List them.
36. Find the number of subsets of size 12, taken from a set of size 15.
37. Suppose your professor wants you to choose 5 exercises from an exercise set containing 55 exercises. In how many ways can you do this?
38. Your professor has a list of 15 questions for the next exam. How many 10 question exams can she make up from this list?
39. How many ways are there to form two teams of 5 people, each from a different group of 10 people? How many ways are there to form two teams of 5 people from one group of 10 people?
40. a) How many 3-tuples are there whose entries are either s or f?
    b) How many of these 3-tuples have exactly two f's?
    c) Check your answers to parts a) and b) by making a complete list of all relevant 3-tuples.
41. a) How many 4-tuples are there whose entries are either 1 or 2?
    b) How many of these 4-tuples have exactly one 1?
    c) Check your answers to parts a) and b) by making a complete list of all relevant 4-tuples.
42. a) How many 6-tuples are there whose entries are either A or B?
    b) How many of these 6-tuples have exactly two A's?
43. a) How many 7-tuples are there whose entries are either H or T?
    b) How many of these 7-tuples have exactly three H's?
44. a) How many 3-tuples are there whose entries are either A, B or C?
    b) How many of these 3-tuples have exactly one A?
45. a) How many 4-tuples are there whose entries are either H, T or N?
    b) How many of these 4-tuples have exactly 2 H's?
46. A man has 5 friends. How many ways can he form a party consisting of *at least* one friend?
47. A married couple decides to have a party. The husband has 8 friends, and the wife has 9 friends. Unfortunately, none of the husband's friends will attend a party that is attended by any of his wife's friends. How many different parties of size 4 can the couple have?
48. A certain club consists of 5 men and 6 women.
    a) How many ways are there to form a committee of 3 people?
    b) How many ways are there to form a committee consisting of 3 men and 4 women?
49. A certain club consists of 5 men and 6 women.
    a) How many ways are there to form a committee of 6 people if a certain pair of women refuse to serve on the same committee? *Hint:* one way

to solve this problem is to count the total number of committees, and then subtract the number for which both women are members.

b) How many ways are there to form a committee of 4 men and 3 women if two of the men refuse to serve on the same committee?

50. A student must choose any 10 questions from a 14 question test.
    a) In how many ways can he do this?
    b) In how many ways can he do this if he must choose 6 questions from the first 8 questions and 4 questions from the last 6 questions?
51. A man has n friends. He is able to invite a different combination of 3 of his friends to his home each night for a full year. What is the smallest possible value of n?
52. Give an algebraic proof of formula (3), which is $C(n, k) = C(n, n-k)$.

Chapter 2
Elementary Probability

2.1 An Introduction to Probability

Now that we have sufficient background in elementary counting techniques, we can consider one of the most important applications of these techniques, namely, to the theory of probability. Probability theory is a branch of mathematics that deals with methods for predicting the likelihood of outcomes of future experiments.

As a simple example, if a coin is perfectly balanced, then we are willing to make the assumption that when the coin is tossed in the air, it is equally likely for it to land with heads facing up as with tails facing up. In a situation such as this, we will say that the *probability* that the coin will land with heads facing up on a future toss is $\frac{1}{2}$. (The same is true for tails.)

This type of statement is typical of probability theory. Of course, we can never expect to predict the outcome of a future experiment with absolute certainty, but we can still obtain very useful information.

Pierre de Fermat
(1601-1665)

Probability theory was first developed, in the late seventeenth and early eighteenth centuries, to deal with questions involving games of chance, and many of our examples will involve experiments such as tossing coins, rolling dice, and drawing cards. The mathematicians Blaise Pascal, whom we encountered in the previous chapter, and Pierre de Fermat, are generally regarded as having founded the theory of probability.

Sample Spaces

The set of possible outcomes of an experiment is called the **sample space** of the experiment. For example, if we toss a coin, then the sample space consists of the two possible outcomes Heads and Tails. For simplicity, we will abbreviate these outcomes as H and T, and so in this case the sample space is the set $S = \{H, T\}$.

Example 1

Describe the sample space for each of the following experiments.

a) Tossing two coins
b) Rolling a single die
c) Rolling a pair of dice

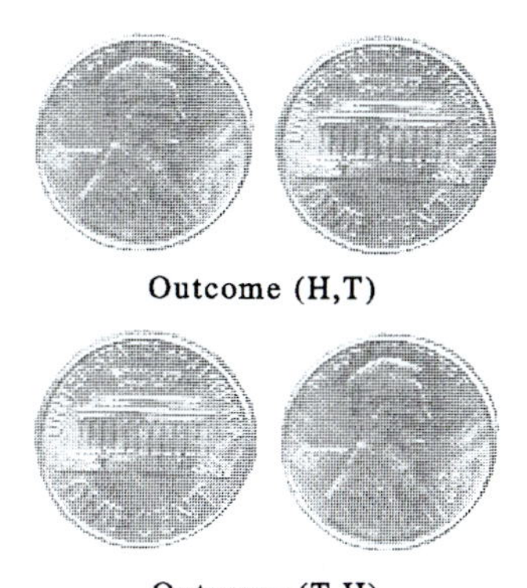

Outcome (H,T)

Outcome (T,H)

d) Drawing 5 cards, order not counting, from an ordinary deck of 52 cards

Solutions

a) If we toss two coins in the air, each possible outcome can be represented by an ordered pair. For example, the ordered pair (H, T) represents the outcome of obtaining a head on the first coin and a tail on the second coin, and the ordered pair (T, H) represents the outcome of obtaining a tail on the first coin and a head on the second. Thus, the sample space for this experiment is the set

$$S = \{(H, H), (H, T), (T, H), (T, T)\}$$

b) If we roll a single die, the possible outcomes are the numbers from 1 to 6, and so the sample space for this experiment is the set

$$S = \{1, 2, 3, 4, 5, 6\}$$

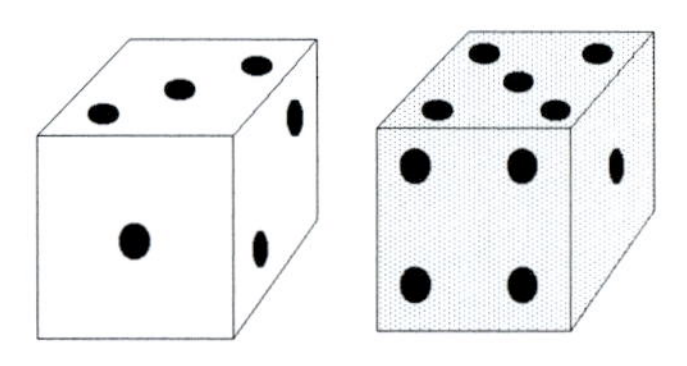

Outcome (3,5)

c) If we roll two dice, then each outcome can be represented as an ordered pair, whose first component gives the value on the first die, and whose second component gives the value on the second die. Thus, for example, the ordered pair $(3, 5)$ represents the outcome of obtaining a 3 on the first die and a 5 on the second die. In this case then, the sample space is the set

$$S = \{(1, 1), (1, 2), (1, 3), \ldots, (6, 4), (6, 5), (6, 6)\}$$

and the fundamental counting principle tells us that there are 36 ordered pairs in this sample space.

Outcome {9♦,8♣,K♠,5♠,A♦}

d) If we draw 5 cards from a deck of cards, then each outcome can be represented as a subset of size 5, taken from the set of 52 cards. Thus, the sample space is the set of all subsets of size 5, taken from the set of 52 cards. We saw in Example 3 of Section 1.3 that there are

$$C(52, 5) = 2,598,960$$

possible hands, and so we won't be writing out the sample space explicitly in this case. ★

In constructing the sample space for a particular experiment, it may help to ask yourself the question, "What information do I need to identify *precisely* any particular outcome?" This should give you a hint as to the form of the sample space.

Events

In dealing with questions relating to probability, we will be interested not just in the probability of a single outcome, but in the probability that any one of a certain subset of outcomes will occur. For example, in playing the game of craps, a player roles two dice, and wins if the *sum* of the numbers on the top faces of the dice is equal to 7. He does not particularly care how this sum occurs. Any outcome that produces a sum of 7 is as good as any other. For instance, the outcome $(1,6)$ is just as good as the outcomes $(2,5),(3,4),(4,3),(5,2)$ or $(6,1)$. This leads us to make the following definition.

Definition

Let S be the sample space of an experiment. Then any subset of S is called an **event.** ★

We will use letters such as E and F to represent events.

Example 2

a) Let S be the sample space associated with the experiment of tossing two coins. Find the event E of getting at least one head.
b) Let S be the sample space associated with the experiment of drawing a card from a deck of cards. Find the event E of getting a heart.
c) Let S be the sample space associated with rolling two dice. Find the event E of getting a sum equal to 7.

Solutions

a) In this case, the sample space is

$$S = \{(H,H),(H,T),(T,H),(T,T)\}$$

and the event E is the subset

$$E = \{(H,H),(H,T),(T,H)\}$$

since these are outcomes that result in at least one head.

b) In this case, the sample space is the set S consisting of all 52 cards, and the event E is the subset of S consisting of all 13 hearts.
c) The sample space for this experiment is the set of all 36 ordered pairs whose components come from the set $\{1,2,3,4,5,6\}$, and the event E is the set

$$E = \{(1,6),(2,5),(3,4),(4,3),(5,2),(6,1)\}$$

since these are the outcomes that result in a sum equal to 7. ★

The Probability of an Outcome

Now we are ready for the main concept in probability theory, namely, the probability of an outcome. Intuitively speaking, the probability of an outcome is intended to represent the *likelihood* that this particular outcome will occur. To get a feeling for what this means, let us consider an example. Suppose we draw a card at random from a thoroughly shuffled deck of 52 cards. How would we express the likelihood (or probability) of drawing a heart?

One possibility would be to imagine that we could repeat this identical experiment a very large number of times, keeping track of the results. Since there are 13 hearts in a total of 52 cards, it seems reasonable that, over a very large number of repetitions, we can expect to get about 13 hearts for every 52 repetitions, that is, we should get a heart about $\frac{13}{52} = \frac{1}{4}$ of the time. We cannot be certain to get a heart *exactly* $\frac{1}{4}$ of the time but, and this is the key point, the more times we repeat the experiment, the closer we should come to getting a total number of hearts equal to $\frac{1}{4}$ the number of repetitions. For this reason, we say that the *probability* of getting a heart on any given trial is $\frac{1}{4}$.

In short, when we say that an outcome has probability equal to, for example, $\frac{1}{4}$, we mean that the more times we repeat the experiment, the more likely we are to get this outcome approximately $\frac{1}{4}$ of the time.

You can see from this discussion that the probability of an outcome is a number between 0 and 1. If an outcome has probability 0, we mean that it can *never* occur. If an outcome has probability 1, then it must *always* occur. When an outcome has a probability that is greater than 0 and less than 1, we cannot make such categorical statements.

Of course, when an experiment is performed, some outcome from the sample space must occur. Therefore, we require that the sum of the probabilities of *all* of the outcomes in the sample space be equal to 1. Let us formalize our discussion in a definition.

Definition

Let $S = \{s_1, \dots, s_n\}$ be the sample space associated with a particular experiment. Thus, $s_1, \dots, s_n$ are the possible outcomes of the experiment. The **probability** of the outcome s_i is a number p_i with the following two properties.

1) $0 \leq p_i \leq 1$

2) $p_1 + p_2 + \cdots + p_n = 1$

The sample space $S = \{s_1, \dots, s_n\}$, together with the probabilities $\{p_1, \dots, p_n\}$, is called a **probability model** of the experiment. ★

In the definition of probability, we said nothing about likelihood. It is customary to define probabilities simply as numbers with the two properties described in the definition, but it is important to keep in mind that probabilities are intended to represent likelihoods.

Given an experiment, how do we determine the probabilities of the various outcomes? If the experiment is very complex, this can be a difficult task. Imagine, for instance, that we have bacterial cultures growing in two different types of media. We sample one bacterium from a culture of each type. The probability that a bacterium from cultures of the first type is larger than a bacterium from a culture of the second type can only be approximated after many repetitions of the experiment. Such probabilities are referred to as **empirical probabilities**. In this book, we will restrict attention to experiments in which we can fairly easily determine the probabilities of each outcome.

Often the probabilities of the outcomes are given to us in the way the experiment is described. For example, in tossing a *fair* coin, it is equally likely for heads to occur as tails. Hence, the probability of getting heads is $\frac{1}{2}$ and the probability of getting tails is also $\frac{1}{2}$. In fact, this is precisely what we mean by a "fair" coin. While there is no such thing as a perfectly fair coin in real life, coins that have not been tampered with are close enough so that we could not detect the difference.

Similarly, when we use the phrase *draw a card at random*, we mean that the card is drawn in such a way that each card is equally likely to be chosen. Hence, the probability of drawing any particular card is $\frac{1}{52}$.

Many experiments are such that the probabilities of all outcomes are the same, say p. When this is the case, if the sample space has size n, then when we add the probability p a total of n times, we should get 1, that is

$$\underbrace{p+p+\cdots+p}_{n \text{ times}} = 1$$

But this is the same as $np = 1$, and so $p = 1/n$. Let us put this fact into a theorem.

Theorem 1

When the outcomes of an experiment are equally likely and the sample space has size n, then the probability of each outcome is $\frac{1}{n}$. ★

Now let us consider some examples.

Example 3

a) A fair coin is tossed in the air. What is the probability of obtaining heads?
b) A coin is weighted in such a way that heads comes up twice as often as tails. What is the probability of getting heads when the coin is tossed?

Solutions

a) Since the coin is fair, the two possible outcomes (heads and tails) are equally likely. Hence, according to Theorem 1, the probability of getting heads is $\frac{1}{2}$.
b) In this case, we cannot apply Theorem 1, since the two outcomes are not equally likely. However, if we denote the probability of getting heads by h and the probability of getting tails by t, then since the two probabilities must sum to 1, we have

$$h + t = 1$$

Also, the fact that heads is twice as likely to occur as tails means that $h = 2t$. Substituting this value of h gives

$$2t + t = 1$$

or

$$3t = 1$$

and so $t = \frac{1}{3}$ and $h = \frac{2}{3}$. Thus, the probability of getting heads is $\frac{2}{3}$. ★

Example 4

Imagine a mini-lotto game in which you choose one number from the numbers 1 through 10. The winning number is determined using a machine that has 10 ping-pong balls, numbered 1 through 10. The balls are blown into the air and whichever ball falls first into a chute is the winning number.

a) If the balls are all identical (except for the numbering), what are the probabilities of each number occurring?

b) Suppose that the balls numbered 9 and 10 are heavier than the other 8 balls, and are twice as likely to fall into the chute. In this case, what are the probabilities of each number occurring?

Solutions

a) Since there are 10 equally likely outcomes in this case, Theorem 1 tells us that the probability of each outcome is $1/10$.

b) Let us denote by p the probability of getting the number 1. Then p is also the probability of getting each of the numbers 2 through 8. The probability of getting each of the numbers 9 and 10 is $2p$, since each of them is twice as likely as each of the numbers 1 through 8. Summing all of these probabilities gives

$$\underbrace{p + \cdots + p}_{8\text{ times}} + 2p + 2p$$

or

$$12p = 1$$

or

$$p = \frac{1}{12}$$

Hence, the probability of getting each of the numbers 1 through 8 is $\frac{1}{12}$ and the probability each of the numbers 9 and 10 is $\frac{2}{12} = \frac{1}{6}$. ★

The Probability of an Event

Often, we are interested in not just the probability of a single outcome, but the probability that any one of the outcomes from a given event will occur. It seems natural to define the probability of an event to be the sum of the probabilities of the outcomes in that event. Let us make that official.

Definition

Let S be the sample space associated with a particular experiment. If E is an event, then the **probability of E**, denoted by $\mathcal{P}(E)$, is the sum of the probabilities of the outcomes in that event. ★

If the probabilities of each outcome are all the same, we can get a simple formula for the probability of an event. Recall from Theorem 1 that when the outcomes of an experiment are equally likely and the sample space has size n, then the probability of each outcome is $\frac{1}{n}$. If an event E has size k (that is, E contains k outcomes), then the probability of that event is given by

$$\mathcal{P}(E) = \underbrace{\frac{1}{n} + \frac{1}{n} + \cdots + \frac{1}{n}}_{k \text{ times}} = \frac{k}{n}$$

This gives us a companion theorem to Theorem 1. We use the notation $|A|$ to denote the size of a set A.

Theorem 2

Let S be a sample space in which each outcome is equally likely to occur and let E be an event in S. Then the probability of E is given by the formula

$$\mathcal{P}(E) = \frac{|E|}{|S|}$$

In words, the probability of E is the number of outcomes in E divided by the total number of outcomes in the sample space S. ★

Here are some additional examples.

Example 5

a) Three fair coins are tossed in the air. What is the probability of the event of getting exactly 2 heads?
b) Two fair dice are rolled. What is the probability of the event of getting a sum of 7?
c) A card is drawn at random from a deck of 52 cards. What is the probability that it is either a heart or an ace?

Solutions

a) The sample space for this experiment is the set S of ordered triples whose components come from the set $\{H, T\}$. For example, the ordered triple (H, T, T) represents the outcome of getting a head on the first coin, and tails on the other two coins. According to Theorem 2 of Section 1.3, there are $2^3 = 8$ such ordered triples, of which $C(3, 2) = 3$ have exactly 2 heads. In fact, the event E of obtaining exactly 2 heads is the subset

$$E = \{(H,H,T),(H,T,H),(T,H,H)\}$$

Thus, the probability of getting exactly 2 heads is

$$\mathcal{P}(\text{exactly 2 heads}) = \mathcal{P}(E) = \frac{3}{8}$$

b) In this case, the sample space has 36 elements in it, and the event E of getting a sum of 7 has 6 elements in it, as we saw in Example 2 part c). Hence, the probability of getting a sum of 7 is

$$\mathcal{P}(\text{sum of } 7) = \mathcal{P}(E) = \frac{6}{36} = \frac{1}{6}$$

c) The sample space for this experiment has size 52. Also, since there are 13 hearts and 3 additional aces that are not hearts, the event E has size $13 + 3 = 16$. Therefore, the probability of getting a heart or an ace is

$$\mathcal{P}(\text{heart or ace}) = \mathcal{P}(E) = \frac{16}{52} = \frac{4}{13}$$ ★

Example 6

Ten fair coins are tossed in the air. What is the probability of getting exactly 4 tails?

Solution

A typical outcome of this experiment can be described as an ordered 10-tuple, whose components are either H or T. For example, the 10-tuple

$$(H,H,H,T,H,T,T,H,T,T)$$

is one possible outcome. We know from Theorem 2 of Section 1.3 that the number of such 10-tuples is

$$2^{10} = 1024$$

and so the sample space S of this experiment has size 1024.

Next, we must compute the size of the event E of getting exactly 4 tails. But a 10-tuple is in E if and only if it has exactly 4 T's, and Theorem 2 of Section 1.3 tells us that there are $C(10,4)$ of these. Hence, there are $C(10,4) = 210$ outcomes in the event E.

Now that we know the sizes of S and E, we can compute the probability of E

$$\mathcal{P}(\text{exactly 4 tails}) = \mathcal{P}(E) = \frac{210}{1024} \approx 0.21$$ ★

Example 7

Studies of the weather in a certain city over the last several decades have shown that, for the month of January, the probability of having a certain number of sunny days is as follows:

$$\begin{aligned}&\mathcal{P}(0\text{–}5 \text{ sunny days}) = 0.65, \quad \mathcal{P}(6\text{–}12 \text{ sunny days}) = 0.2,\\ &\mathcal{P}(13\text{–}19 \text{ sunny days}) = 0.1, \quad \mathcal{P}(20\text{–}25 \text{ sunny days}) = 0.04,\\ &\mathcal{P}(26\text{–}31 \text{ sunny days}) = 0.01\end{aligned}$$

What is the probability of having at least 13 sunny days? What is the probability of having at most 5 sunny days or at least 26 sunny days?

Solution The event of having at least 13 sunny days consists of the three outcomes: 13–19 sunny days, 20–25 sunny days and 26–31 sunny days. Adding the probabilities of these three outcomes gives

$$\begin{aligned}&\mathcal{P}(\text{at least 13 sunny days})\\ &= \mathcal{P}(13\text{–}19 \text{ sunny days}) + \mathcal{P}(20\text{–}25 \text{ sunny days}) + \mathcal{P}(26\text{–}31 \text{ sunny days})\\ &= 0.1 + 0.04 + 0.01 = 0.15\end{aligned}$$

Similarly,

$$\begin{aligned}&\mathcal{P}(\text{at most 5 sunny days or at least 26 sunny days})\\ &= \mathcal{P}(0\text{–}5 \text{ sunny days}) + \mathcal{P}(26\text{–}31 \text{ sunny days})\\ &= 0.65 + 0.01 = 0.66\end{aligned}$$

★

Let us conclude this section with a final comment on sample spaces. The probabilities that we determine for a given experiment depend on what we consider to be the outcomes of the experiment. Let us illustrate this with an example.

Example 8

Consider the experiment of rolling two fair dice. If we record the values on each die, the sample space consists of the 36 ordered pairs

$$S = \{(1,1), (1,2), (1,3), \ldots, (6,4), (6,5), (6,6)\}$$

as we saw in Example 1. Since the dice are fair, each ordered pair is equally likely to occur and so the probability of each outcome is $\frac{1}{36}$.

However, for some purposes, we may be interested only in the *sum* of the two numbers on the dice. If we record only this sum as our outcome, the sample space consists of the possible sums, which are the numbers 2 through 12. That is, the sample space is

$$T = \{2, 3, \ldots, 12\}$$

In this case, the outcomes are *not* equally likely. To determine the probabilities of each outcome, we note that each outcome in T corresponds to an *event* in the sample space S.

For instance, a sum of 3 occurs when either of the ordered pairs $(1, 2)$ or $(2, 1)$ occurs and so the outcome 3 in T corresponds to the event

$$E = \{(1, 2), (2, 1)\}$$

in S. Hence, the probability of getting a sum of 3 is the probability of the event E,

$$\mathcal{P}(\text{sum equals } 3) = \mathcal{P}(E) = \frac{2}{36} = \frac{1}{18}$$

As another example, a sum of 7 corresponds to the event

$$F = \{(1, 6), (2, 5), (3, 4), (4, 3), (5, 2), (6, 1)\}$$

in S. Since $|F| = 6$, we have

$$\mathcal{P}(\text{sum equals } 7) = \mathcal{P}(F) = \frac{6}{36} = \frac{1}{6}$$ ★

EXERCISES

Define or discuss each of the following terms.

a) Probability Theory
b) Outcome
c) Sample space
d) Event
e) Probability of an event
f) The law of large numbers

1. A card is drawn at random from a deck of 52 cards.
   a) Find the probability of drawing a king.
   b) Find the probability of drawing a diamond.
   c) Find the probability of drawing a red card.
2. A card is drawn at random from a deck of 52 cards.
   a) Find the probability of drawing a face card. (A *face card* is a king, queen or jack.)
   b) Find the probability of drawing either a king or a queen.
   c) Find the probability of drawing either a red card with face value at least 10 or a black card with face value at most 6.
3. A single fair die is rolled.
   a) Find the probability of obtaining a 6.
   b) Find the probability of obtaining a number greater than or equal to 4.
   c) Find the probability of obtaining a number greater than 4 or less than 2.
4. A pair of fair dice are rolled.
   a) Find the probability that the sum is equal to 5.
   b) Find the probability that the sum is equal to 11.
   c) Find the probability that the sum is at most 4.
5. A coin is tampered with so that tails is 3 times more likely to occur than heads.
   a) What is the probability of getting heads?
   b) What is the probability of getting tails?
6. A die has six sides, but two sides have only 1 dot. The other four sides have $2, 3, 4$ and 5 dots. Assume that each side is equally likely to occur.
   a) What is the probability of getting a 1?

   b) What is the probability of getting a 2?
   c) What is the probability of getting an even number?
   d) What is the probability of getting a number less than 3?
7. Consider a lotto game as described in Example 4. Assume that the balls numbered 8, 9 and 10 are heavier than the other balls, and are each twice as likely to fall into the chute as each of the others. Find the probability of getting each of the numbers 1 through 10. What is the probability of getting an even number?
8. Consider a lotto game as described in Example 4. Assume that the balls numbered 1 through 7 all equally likely to fall into the chute, the balls numbered 8 and 9 are twice as likely to fall into the chute as ball 1 and the ball numbered 10 is 3 times as likely to fall into the chute as ball 1. Find the probability of getting each of the numbers 1 through 10. What is the probability of getting an even number?
9. A pair of fair dice are rolled. Find the probability of getting a sum that is even.
10. Three fair dice are rolled. Find the probability of getting exactly one 6.
11. Four fair coins are tossed. Find the probability of getting exactly 2 heads.
12. Four fair coins are tossed. Find the probability of getting at least 2 heads.
13. A basket contains 5 red balls, 3 black balls, and 4 white balls. A ball is chosen at random from the basket.
    a) Find the probability of choosing a red ball.
    b) Find the probability of choosing a white ball or a red ball.
    c) Find the probability of choosing a ball that is not red.
14. A certain true and false test contains 10 questions. A student guesses randomly at each question.
    a) What is the probability that he will get all 10 questions correct?
    b) What is the probability that he will get at least 9 questions correct?
    c) What is the probability that he will get at least 8 questions correct?
15. Consider an urn that contains 5 red balls, 6 blue balls and 10 black balls. If you reach into the urn, select a ball at random (without looking) and record only the color. What is the sample space and what are the probabilities of each outcome? What is the probability that the ball is not red?
16. An American roulette wheel has slots numbered 00, 0 and 1-36. The numbers 0 and 00 are green and the other numbers are half red and half black. Neither 0 nor 00 are considered even or odd. Assuming that the wheel is perfectly balanced, find the probability of the following
    a) getting the number 00
    b) getting a red number
    c) getting an even number

    Suppose that the wheel is altered so that 0 and 00 come up twice as often as they would normally. What is the probability of getting an odd number?
17. Consider the dart board shown below

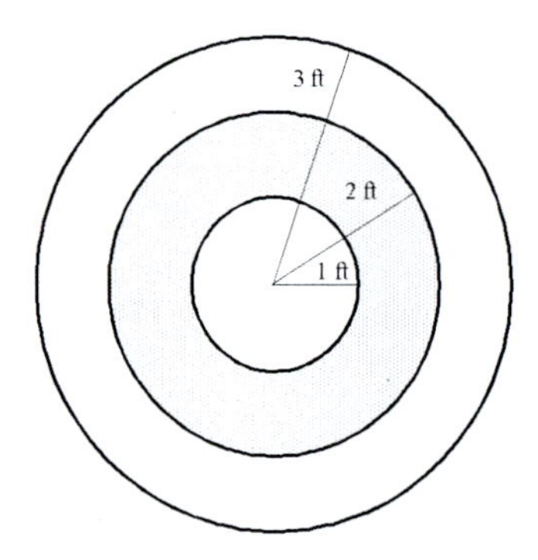

If a dart is equally likely to hit the board at any location, the probability that it lands in a given region is equal to the area of that region divided by the entire area of the board. Using the formula $A = \pi r^2$ for the area of a circle of radius r, compute the probabilities of hitting each of the three regions.

18. Consider the game of chance shown below

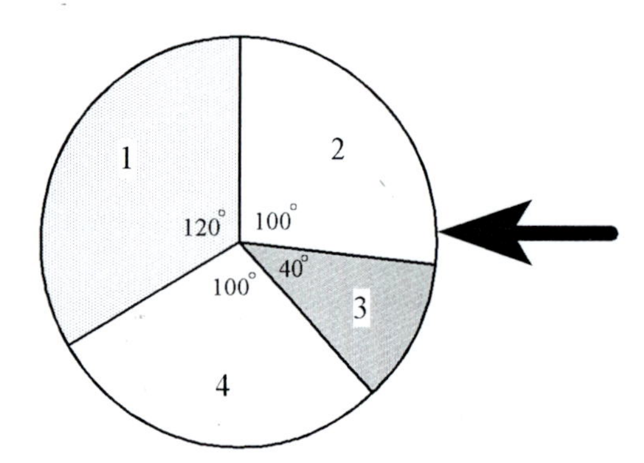

The wheel is spun and the number of the region that stops next to the pointer is recorded as the outcome. The angles of each region are given in the figure. Can you figure out the probabilities of getting a 1, 2, 3 and 4? Do you need to know the radius of the wheel? What is the probability of the spinner landing in one of the two darker regions? What is the probability of getting an even numbered region?

Exercises 19-24 refer to an experiment in which 5 *cards are drawn at random from a deck of* 52 *cards.*

19. Find the probability of drawing 4 aces. *Hint:* how many 5–card hands contain 4 aces?
20. Find the probability of drawing 2 queens and 3 kings. *Hint:* how many 5–card hands have 2 queens and 3 kings? (There are 4 Queens and 4 Kings in a deck of cards.)
21. Find the probability of drawing 4 of a kind (that is 4 aces, or 4 kings, or 4 queens, and so on.)
22. Find the probability of drawing at least 1 ace. *Hint:* there are $C(48, 5)$ hands that contain no aces.

23. Find the probability of drawing a spade flush, that is, 5 spades.
24. Find the probability of drawing a royal flush. (A royal flush consists of the Ace, King, Queen, Jack and 10 of the same suit.)

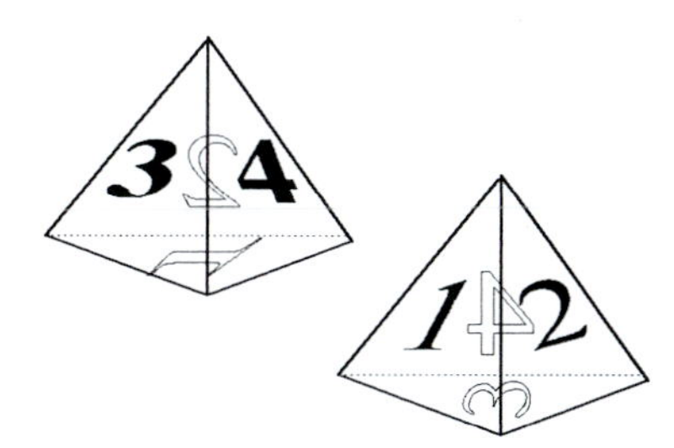

Exercises 25-29 refer to the two dice shown shown above, which are tetrahedra, whose faces are numbered 1, 2, 3 and 4.

25. What is the sample space for the experiment of rolling two tetrahedral dice? How large is the sample space?
26. If a single tetrahedral die is rolled, what is the probability that the face down side is a 3?
27. If a single tetrahedral die is rolled, what is the probability that the face down side is not a 1?
28. If two tetrahedral dice are rolled, what is the probability that the sum of the face down sides is 5?
29. If two tetrahedral dice are rolled, what is the probability that the sum of the face down sides is even?
30. Consider a roulette wheel, as described in Exercise 16. Suppose we change the numbers on the wheel by dividing all even numbers by 2 and adding 1. What are the new numbers on the wheel? What is the probability of getting an odd number?
31. A fair die is rolled and a card is chosen at random. What is the probability that the number on the die matches the number on the card? (An ace is counted as a one.)
32. Studies of the weather in a certain city over the last several decades have shown that, for the month of March, the probability of having a certain amount of sun/smog is as follows:

$\mathcal{P}(\text{full sun/no smog}) = 0.07$ $\mathcal{P}(\text{full sun/light smog}) = 0.09$,
$\mathcal{P}(\text{full sun/heavy smog}) = 0.12$, $\mathcal{P}(\text{haze/no smog}) = 0.09$,
$\mathcal{P}(\text{haze/light smog}) = 0.07$, $\mathcal{P}(\text{haze/heavy smog}) = 0.11$
$\mathcal{P}(\text{no sun/no smog}) = 0.16$, $\mathcal{P}(\text{no sun/light smog}) = 0.12$
$\mathcal{P}(\text{no sun/heavy smog}) = 0.17$,

What is the probability of having at a fully sunny day? What is the probability of having at day with some sun? What is the probability of having a day with no or light smog?

2.2 Independent Events; Binomial Probability

Now that we have completed our basic introduction to probability, we can discuss some of the key ideas in the field.

The Complement of an Event

If E is an event, then the **complement** of E is simply the event consisting of all outcomes in the sample space that are *not* in E. We denote the complement of E by E^c. There is a simple relationship between the probability of an event E and the probability of its complement E^c. Since all outcomes in the sample space must be in either E or E^c and since no outcome can be in both, when we compute the sum

$$\mathcal{P}(E) + \mathcal{P}(E^c)$$

we are simply computing the sum of the probabilities of all possible outcomes, and so we must get 1. Hence

$$\mathcal{P}(E) + \mathcal{P}(E^c) = 1$$

This can be solved for either $\mathcal{P}(E)$ or $\mathcal{P}(E^c)$ to give two very useful formulas

$$\mathcal{P}(E^c) = 1 - \mathcal{P}(E) \tag{1}$$

$$\mathcal{P}(E) = 1 - \mathcal{P}(E^c) \tag{2}$$

These formulas are useful, since it is sometimes the case that computing one of the probabilities $\mathcal{P}(E)$ or $\mathcal{P}(E^c)$ is much easier than computing the other. Here is an example.

Example 1

Two cards are drawn at random from a deck of 52 cards, each one being replaced before the next is drawn. What is the probability of drawing at least one spade?

Solution The sample space S is the set of all ordered pairs (x, y), where x is the first card and y is the second card drawn. Thus, S has size

$$|\, S \,| = 52 \times 52 = 2704$$

Let E be the event of getting at least one spade. We could compute the probability of E directly, but it is easier to compute the probability of E^c, which is the event that *no* spades were drawn. (The complement of *at least one* is *none*.) After all, there are $52 - 13 = 39$ cards that are not spades, and so there are

$$39 \times 39 = 1521$$

ordered pairs with no spades, that is

$$|\, E^c \,| = 1521$$

Hence, the probability of E^c is

$$\mathcal{P}(E^c) = \frac{1521}{2704}$$

and so formula (2) gives

$$\mathcal{P}(E) = 1 - \mathcal{P}(E^c) = 1 - \frac{1521}{2704} = \frac{1183}{2704} = 0.4375$$ ★

The Union and Intersection of Events

If E and F are events from the same sample space, we often want to compute the probability that *at least one* of the events E or F has occurred, or the probability that *both* of the events E and F have occurred. As you may know, the **union** $E \cup F$ is the set of outcomes that are in E *or* F (or both), and the **intersection** $E \cap F$ is the set of outcomes that are in *both* E and F.

Example 2

Suppose we draw a card from a deck of cards. Let E be the event of drawing a spade and let F be the event of drawing a face card. Thus,

$$E = \{A\spadesuit, 2\spadesuit, 3\spadesuit, \ldots, J\spadesuit, Q\spadesuit, K\spadesuit\}$$

and

$$F = \{J\clubsuit, J\diamondsuit, J\heartsuit, J\spadesuit, Q\clubsuit, Q\diamondsuit, Q\heartsuit, Q\spadesuit, K\clubsuit, K\diamondsuit, K\heartsuit, K\spadesuit\}$$

Then $E \cup F$ is the event of drawing either a spade *or* a face card, that is,

$$E \cup F = \{A\spadesuit, 2\spadesuit, 3\spadesuit, \ldots, J\spadesuit, Q\spadesuit, K\spadesuit, J\clubsuit, J\diamondsuit, J\heartsuit, Q\clubsuit, Q\diamondsuit, Q\heartsuit, K\clubsuit, K\diamondsuit, K\heartsuit\}$$

Also, $E \cap F$ is the event of drawing both a spade *and* a face card, that is,

$$E \cap F = \{J\spadesuit, Q\spadesuit, K\spadesuit\}$$ ★

If E and F are events from the same sample space, and if E and F have no outcomes in common, we say that they are **mutually exclusive**. In symbols, the events E and F are mutually exclusive if the intersection $E \cap F$ is the empty set, that is,

$$E \cap F = \emptyset$$

Example 3

The event E of drawing a spade and the event H of drawing a heart from a deck of cards are mutually exclusive, since no outcome can be both a spade and a heart. That is, $E \cap H = \emptyset$.

On the other hand, the event E of drawing a spade and the event F of drawing a face card are not mutually exclusive, since the outcomes $J\spadesuit$, $Q\spadesuit$ and $K\spadesuit$ are in both events. In symbols,

$$E \cap F = \{J\spadesuit, Q\spadesuit, K\spadesuit\}$$

★

When two events E and F are mutually exclusive, then the probability of their union $E \cup F$ is just the sum of the probabilities of each event, in symbols

$$\mathcal{P}(E \cup F) = \mathcal{P}(E) + \mathcal{P}(F)$$

However, when the events are not mutually exclusive, this formula does not hold. For instance, if E is the event of drawing a spade from a deck of cards, the probability of E is (as we have seen)

$$\mathcal{P}(E) = \frac{13}{52} = \frac{1}{4}$$

If F is the event of drawing an ace, the probability of F is

$$\mathcal{P}(F) = \frac{4}{52} = \frac{1}{13}$$

Now, the union $E \cup F$ consists of all cards that are either spades or aces, and there are 16 such cards – the 13 spades plus the 3 additional aces of the other suits (hearts, diamonds and clubs). Hence, the probability of the union is

$$\mathcal{P}(E \cup F) = \frac{16}{52}$$

and this is *not* equal to the sum

$$\mathcal{P}(E) + \mathcal{P}(F) = \frac{1}{4} + \frac{1}{13} = \frac{17}{52}$$

The reason that these two quantities are not equal is that, in the sum $\mathcal{P}(E) + \mathcal{P}(F)$, we are counting the ace of spades twice – once in $\mathcal{P}(E)$ since the ace of spades is a spade, and once in $\mathcal{P}(F)$ since the ace of spades is an ace. However, in the probability $\mathcal{P}(E \cup F)$, this card is only counted once. Notice that the *intersection* of the two events E and F is the set containing precisely the ace of spades, which is the card that was counted twice in $\mathcal{P}(E) + \mathcal{P}(F)$. To count it only once, we subtract, to get

$$\mathcal{P}(E \cup F) = \mathcal{P}(E) + \mathcal{P}(F) - \mathcal{P}(E \cap F) \tag{3}$$

As a check, we have

$$\mathcal{P}(E \cap F) = \mathcal{P}(\text{Ace of spades}) = \frac{1}{52}$$

and

$$\mathcal{P}(E) + \mathcal{P}(F) - \mathcal{P}(E \cap F) = \frac{1}{4} + \frac{1}{13} - \frac{1}{52} = \frac{16}{52} = \mathcal{P}(E \cup F)$$

and so we see that formula (3) is true in this case.

Formula (3) holds for all sample spaces and all events E and F and so we have a nice (and extremely important) theorem.

Theorem 1

If E and F are events in the same sample space S, then the probability of the union $E \cup F$ is given by the formula

$$\mathcal{P}(E \cup F) = \mathcal{P}(E) + \mathcal{P}(F) - \mathcal{P}(E \cap F) \tag{4}$$

If E and F are mutually exclusive, then $E \cap F = \emptyset$ and so $\mathcal{P}(E \cap F) = 0$ and this formula becomes

$$\mathcal{P}(E \cup F) = \mathcal{P}(E) + \mathcal{P}(F)$$ ★

Formula (4) in Theorem 1 is sometimes called the **Principle of Inclusion and Exclusion**. We can justify formula (4) as follows. In the sum $\mathcal{P}(E) + \mathcal{P}(F)$, we count the probability of each outcome in E and the probability of each outcome in F, that is, we count the probability of each outcome in the union $E \cup F$. However, those outcomes that lie in *both* events E and F get counted twice – once in the number $\mathcal{P}(E)$ and once in the number $\mathcal{P}(F)$. In other words, those outcomes in the intersection $E \cap F$ get counted twice in the sum $\mathcal{P}(E) + \mathcal{P}(F)$. To correct this over count, we must subtract $\mathcal{P}(E \cap F)$, and so the quantity

$$\mathcal{P}(E) + \mathcal{P}(F) - \mathcal{P}(E \cap F)$$

counts each outcome in $E \cup F$ exactly once. Thus

$$\mathcal{P}(E) + \mathcal{P}(F) - \mathcal{P}(E \cap F) = \mathcal{P}(E \cup F)$$

which is formula (4).

Example 4

If a single card is drawn at random from a deck of cards, what is the probability that the card is either a spade or a heart?

Solution The event of getting a spade or a heart is the event $E \cup F$, where E is the event of getting a spade and F is the event of getting a heart. Since these events are mutually exclusive, we have

$$\mathcal{P}(E \cup F) = \mathcal{P}(E) + \mathcal{P}(F) = \frac{13}{52} + \frac{13}{52} = \frac{26}{52} = \frac{1}{2}$$ ★

Example 5

A single card is drawn at random from a deck of cards. What is the probability that it is a red card or a face card?

Solution Let E be the event of drawing a red card and let F be the event of drawing a face card. Since there are 26 red cards and 12 face cards,

$$\mathcal{P}(E) = \frac{26}{52} \text{ and } \mathcal{P}(F) = \frac{12}{52}$$

The intersection $E \cap F$ is the event of drawing a card that is *both* red and a face card, that is, a red face card. Since there are 6 red face cards (Jack, Queen, King of hearts and of diamonds) we have

$$\mathcal{P}(E \cap F) = \frac{6}{52}$$

Hence, Theorem 1 gives

$$\begin{aligned}\mathcal{P}(\text{red card or face card}) &= \mathcal{P}(E \cup F) \\ &= \mathcal{P}(E) + \mathcal{P}(F) - \mathcal{P}(E \cap F) \\ &= \frac{26}{52} + \frac{12}{52} - \frac{6}{52} = \frac{32}{52} = \frac{8}{13}\end{aligned}$$ ★

Example 6

A pair of fair dice is rolled. What is the probability that at least one of the dice is a 1 or that the sum of the two dice is even?

Solution Let E be the event of getting a 1 on at least one of the dice and let F be the event that the sum on the two dice is even. Then E contains 6 ordered pairs of the form $(1, x)$, where x is any number from 1-to 6 and E also contains an additional 5 ordered pairs of the form $(y, 1)$ where y is any number from 2 through 6. We do not include $y = 1$ since the ordered pair $(1, 1)$ was already counted as one of the 6 pairs $(1, x)$. Hence

$$|E| = 6 + 5 = 11$$

To compute the size of the event F, first we remark that in order for the sum of two numbers to be even, both numbers must be even or both numbers must be odd. Since there are 3 even numbers between 1 and 6, the number of ordered pairs (x, y) where both x and y are even is $3 \times 3 = 9$. Similarly, there are 3 odd numbers between 1 and 6 and so the number of ordered pairs (x, y) where both x and y are odd is $3 \times 3 = 9$. Hence

$$|F| = 9 + 9 = 18$$

Now consider the intersection $E \cap F$, which is the event of getting at least one 1 *and* an even sum. If the first die is a 1, then the other die must be an odd number in order for the sum to be even. Hence, the possibilities are

$$(1,1),(1,3) \text{ and } (1,5)$$

If the second number is a 1, the first number must be odd and so we get the additional possibilities

$$(3,1) \text{ and } (5,1)$$

(We don't want to count $(1,1)$ twice.) Hence

$$|E \cap F| = 5$$

Since the sample space has size 36, we can compute the various probabilities

$$\mathcal{P}(E) = \frac{11}{36}, \mathcal{P}(F) = \frac{18}{36}, \mathcal{P}(E \cap F) = \frac{5}{36}$$

Therefore, according to Theorem 1, we have

$$\begin{aligned}\mathcal{P}(E \cup F) &= \mathcal{P}(E) + \mathcal{P}(F) - \mathcal{P}(E \cap F) \\ &= \frac{11}{36} + \frac{18}{36} - \frac{5}{36} = \frac{24}{36} = \frac{2}{3}\end{aligned}$$

That is, the probability of getting at least one 1 or an even sum is $2/3$. ★

Independent Events

Now let us turn to a different topic. Suppose we toss a *fair* coin 99 times, and that we get heads each time. Would you be willing to bet that the 100-th toss will result in another heads? Many people would not, reasoning (incorrectly) that since heads has occurred so many times in a row, an outcome of tails is way overdue.

The fact is, however, that the outcome of each toss of the coin is *independent* of the outcomes of the other tosses, and so the probability of getting a heads on the 100-th toss is still $\frac{1}{2}$, despite the previous results. In other words, even if 99 heads occur in a row, it is just as likely that the 100-th toss is a heads as a tails.

Perhaps the reason for confusion on this point has to do with the probability of getting 100 heads in a row, which is certainly very small. But that is not the same as the probability of getting a heads on the 100-th toss, *given* that the previous 99 tosses have already resulted in heads.

Intuitively speaking, two events E and F are said to be *independent* if the knowledge that one of these events has occurred has no effect on the *probability* of the other event occurring.

As another example of independence, if we draw two cards from a deck of cards, replacing the first card before drawing the second, then the event of drawing a spade on the first card and the event of drawing a spade on the second card are *independent*, since the outcome of the first draw has no effect on the second draw. On the other hand, if we do *not* replace the first card before drawing the second, then the two outcomes are *dependent*, since the outcome of the first draw effects the possibilities for the second draw. (If we draw a spade on the first draw, and do not replace it, we reduce the probability of drawing a spade on the second draw.)

Let us state a formal definition of independent events, which we will motivate with a subsequent example.

Definition

Let E and F be events from the same sample space. Then E and F are **independent** if the probability that both events occur is the product of the probabilities of each event. In symbols,

$$\mathcal{P}(E \cap F) = \mathcal{P}(E) \cdot \mathcal{P}(F)$$ ★

In order to get a feel for why this definition is appropriate, let us return to the example of drawing two cards from a deck of cards.

Example 7

Suppose we draw two cards from a deck of cards, replacing the first before drawing the second. Let E be the event of drawing a spade on the first draw, and let F be the event of drawing a spade on the second draw. Since for each draw, there are 13 spades out of 52 cards, the probabilities of E and F are

$$\mathcal{P}(E) = \mathcal{P}(F) = \frac{13}{52} = \frac{1}{4}$$

Let us compute the probability of drawing a spade on *both* draws, that is, the probability of $E \cap F$. In this case, the sample space S is the set of all ordered pairs of cards, and according to Theorem 2 of Section 1.3, $|S| = 52^2$. The event $E \cap F$ of getting two spades is the set of ordered pairs of spades, and so $|E \cap F| = 13^2$. Hence, the probability of getting two spades is

$$\mathcal{P}(E \cap F) = \frac{13^2}{52^2} = \left(\frac{13}{52}\right)^2 = \left(\frac{1}{4}\right)^2 = \frac{1}{16} = \mathcal{P}(E) \cdot \mathcal{P}(F)$$

Thus, according to the definition, the events E and F are independent, as we would expect. ★

The definition of independent events can be extended to more than two events, although the definition becomes a bit more involved then you might think at first. For instance, three events E, F and G are **independent** if the following two conditions hold.

1) The probability of the intersection $E \cap F \cap G$ is equal to the product of the probabilities, in symbols,

$$\mathcal{P}(E \cap F \cap G) = \mathcal{P}(E) \cdot \mathcal{P}(F) \cdot \mathcal{P}(G)$$

2) Each pair of events from among E, F and G is independent, that is,

a) E and F are independent, in symbols,

$$\mathcal{P}(E \cap F) = \mathcal{P}(E) \cdot \mathcal{P}(F)$$

b) E and G are independent, in symbols,

$$\mathcal{P}(E \cap G) = \mathcal{P}(E) \cdot \mathcal{P}(G)$$

c) F and G are independent, in symbols,

$$\mathcal{P}(F \cap G) = \mathcal{P}(F) \cdot \mathcal{P}(G)$$

As we will see in the upcoming examples, it is common practice to decide whether or not given events are independent based on our intuitive feeling about the events. If the two events E and F are independent, then we can use the formula $\mathcal{P}(E \cap F) = \mathcal{P}(E) \cdot \mathcal{P}(F)$ to compute the probability that both events have occurred. Similarly, if three events E, F and G are independent, we may use the formula

$$\mathcal{P}(E \cap F \cap G) = \mathcal{P}(E) \cdot \mathcal{P}(F) \cdot \mathcal{P}(G)$$

which is a consequence of being independent, to compute the probability that all three events have occurred. More generally, if the n events

$$E_1, \ldots , E_n$$

are independent, then we may use the formula

$$\mathcal{P}(E_1 \cap \cdots \cap E_n) = \mathcal{P}(E_1) \cdots \mathcal{P}(E_n) \tag{5}$$

which is a consequence of being independent, to compute the probability that all n events have occurred.

Example 8

A fair coin is flipped 5 times. What is the probability that all 5 outcomes are heads?

Solution

We can approach this problem in two ways. Using the basic results of the previous section, we would reason as follows. The sample space S for this experiment is the set of all 5-tuples of H's and T's. For instance, one possible outcome is (H, H, T, H, T). According to Theorem 2 of Section 1.3, the size of this sample space is

$$| S | = 2^5 = 32$$

Now, the event we are interested in is $E = \{(H, H, H, H, H)\}$, and its size is $|E| = 1$. Hence, the probability of E is

$$\mathcal{P}(\text{all heads}) = \mathcal{P}(E) = \frac{|E|}{|S|} = \frac{1}{32}$$

The other approach to solving this problem is to observe that the individual tosses of the coin are independent, and so, according to formula (5), since the probability of a single head is $\frac{1}{2}$, the probability of 5 heads is the product

$$\frac{1}{2} \times \frac{1}{2} \times \frac{1}{2} \times \frac{1}{2} \times \frac{1}{2} = \left(\frac{1}{2}\right)^5 = \frac{1}{32} \qquad \bigstar$$

Example 9

A couple has decided to have 3 children. Based on their family histories, they have decided that the probability of having a boy is $\frac{2}{5}$, and the probability of having a girl is $\frac{3}{5}$.

a) What is the probability that the couple will have 3 boys?
b) What is the probability that the couple will have children in the order boy-girl-boy?
c) What is the probability that the couple will have 2 boys and a girl?

Solutions

We begin by noting that having a child has no effect on the genes of the parents, and so has no effect on the gender of subsequent children. Hence, it seems reasonable to assume that the events of having children are independent.

a) Since the probability of having a single boy is $\frac{2}{5}$, the probability of having 3 boys is

$$\mathcal{P}(3 \text{ boys}) = \left(\frac{2}{5}\right)^3 = \frac{8}{125} = 0.064$$

b) Since the events of having children are independent, we have

$$\begin{aligned}\mathcal{P}(\text{boy-girl-boy}) &= \mathcal{P}(\text{boy}) \times \mathcal{P}(\text{girl}) \times \mathcal{P}(\text{boy}) \\ &= \frac{2}{5} \times \frac{3}{5} \times \frac{2}{5} = \frac{12}{125} = 0.096\end{aligned}$$

c) There are $C(3, 1) = 3$ ways in which the couple can have 2 boys and 1 girl, namely,

boy-boy-girl
boy-girl-boy
girl-boy-boy

As in part b), we can easily compute that each of these possibilities has probability $\frac{12}{125}$, the probability of one of these happening is the sum

$$\mathcal{P}(2 \text{ boys, } 1 \text{ girl}) = \frac{12}{125} + \frac{12}{125} + \frac{12}{125} = \frac{36}{125} = 0.288 \qquad \bigstar$$

Now let us have an example combining the concept of independent events and the principle of inclusion and exclusion.

Figure 1

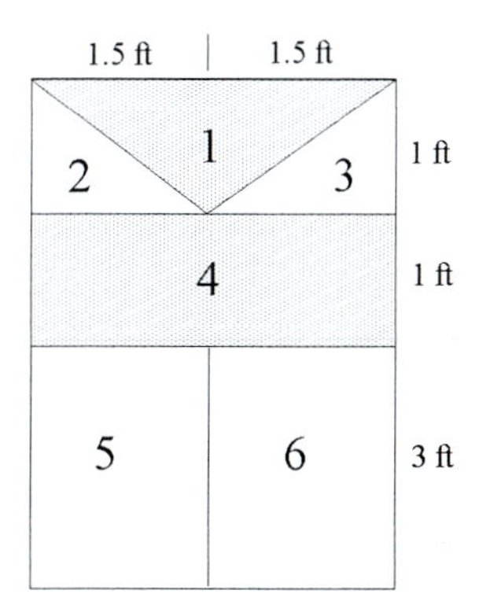

Example 10

Consider the game of shuffleboard, in which a round puck is pushed over a painted region on the ground, such as that shown in Figure 1. Assume that the center of any puck is equally likely to stop at any location inside the rectangular region, that is, assume that the probability that the center of the puck lands in a given region is proportional to the area of that region. We also assume that when more than one puck is pushed, the pucks do not interfere with each other. (For example, each puck may be removed before the next puck is pushed.) If two pucks are pushed, one at a time, what is the probability that the center of at least one puck lands in region 4?

Solution Let E be the event that the first puck lands in region 4, and let F be the event that the second puck lands in region 4. Then we seek the probability of the union $E \cup F$, since this is the event that at least one of the pucks lands in region 4. The probability that a single puck lands in region 4 is the area of region 4, divided by the total area of the rectangle, and so

$$\mathcal{P}(E) = \mathcal{P}(F) = \frac{3 \text{ ft} \times 1 \text{ ft}}{3 \text{ ft} \times 5 \text{ ft}} = \frac{1}{5}$$

The intersection $E \cap F$ is the event that both pucks land in region 4. Our assumption about the pucks not interfering with each other amounts to saying that the events E and F are independent, and so

$$\mathcal{P}(E \cap F) = \mathcal{P}(E)\mathcal{P}(F) = \frac{1}{5} \times \frac{1}{5} = \frac{1}{25}$$

We can now apply formula (4) in Theorem 1 to get

$$\begin{aligned} \mathcal{P}(\text{at least one puck lands in region 4}) &= \mathcal{P}(E \cup F) \\ &= \mathcal{P}(E) + \mathcal{P}(F) - \mathcal{P}(E \cap F) \\ &= \frac{1}{5} + \frac{1}{5} - \frac{1}{25} = \frac{9}{25} \end{aligned}$$

★

Binomial Probabilities

The simplest type of meaningful experiment is one that has only two outcomes. Such experiments are referred to as **Bernoulli experiments**, or **Bernoulli trials**. The two outcomes are often described by the terms *success* and *failure*, and the probability of success is usually denoted by p. Hence, the probability of failure is $1 - p$.

Example 11

The experiment of tossing a coin is a Bernoulli experiment, where we consider heads as a success, and tails as a failure. If the coin is fair, then the probability of success is $p = \frac{1}{2}$.

If we draw a card from a deck of cards, and consider getting an ace as a success and getting any other card as a failure, then we have a Bernoulli experiment. In this case, the probability of success is

$$p = \frac{4}{52} = \frac{1}{13}$$ ★

When *the same* Bernoulli experiment is repeated several times, we refer to the combined experiment as a **binomial experiment**.

Example 12

Tossing a coin 2 times is a binomial experiment. Drawing a card (as described in the previous example) 5 times is a binomial experiment *provided that we replace each card before drawing the next card.* This is necessary since the term binomial experiment refers to repeating the *same* experiment each time. If we do not replace each card, then the set of possible outcomes changes each time, and so we would not be repeating the same experiment. ★

Because the individual Bernoulli trials in a binomial experiment are independent, it is easy to compute the probability of any particular outcome of the binomial experiment, as the following example illustrates.

Example 13

Consider the binomial experiment of rolling a die 5 times, where obtaining a 1 on a given roll is considered a success. Each outcome of this experiment has the form of a 5-tuple of s's (for success) and f's (for failure). For instance, if we roll a 1 on the third and fifth trial, then the outcome of the binomial experiment is the 5-tuple (f, f, s, f, s).

To compute the probability of this outcome, we observe that each of the 5 trials is independent of the others, and since the probability of success is $p = \frac{1}{6}$, and the probability of failure is $1 - p = \frac{5}{6}$, we have, according to the definition of independent events,

$$\begin{aligned}\mathcal{P}(f,f,s,f,s) &= (1-p)\cdot(1-p)\cdot p\cdot(1-p)\cdot p\\ &= \frac{5}{6}\cdot\frac{5}{6}\cdot\frac{1}{6}\cdot\frac{5}{6}\cdot\frac{1}{6} = \frac{125}{7776} \approx 0.016\end{aligned}$$

which is quite small.

We can also compute the probability of obtaining exactly two 1's (that is, of having exactly 2 successes) in any order. The first step is to observe that there are $C(5,2) = 10$ 5-tuples of s's and f's that contain exactly two s's. (This follows from Theorem 2 of Section 1.3.) Furthermore, each of these has probability 0.016 of occurring, and so the probability of getting exactly two 1's is approximately

$$(10)\cdot(0.016) = 0.16$$

★

The results of the previous example are very useful, and we generalize them in the following theorem.

Theorem 2

Consider a binomial experiment that consists of repeating a Bernoulli experiment n times. Let the probability of success in the Bernoulli experiment be denoted by p. Hence, the probability of failure is $1-p$. Then the probability of obtaining exactly k successes in the binomial experiment is

$$C(n,k)p^k(1-p)^{n-k}$$

Proof. The first step is to note that the probability of getting k success and $n-k$ failures *in any given order* is simply the product of k p's and $n-k$ $(1-p)$'s. For instance, the probability of getting k successes, followed by $n-k$ failures, is

$$\underbrace{p\cdots p}_{k}\cdot\underbrace{(1-p)\cdots(1-p)}_{n-k} = p^k(1-p)^{n-k}$$

As another illustration, the probability of getting $n-k$ failures, followed by k successes, is

$$\underbrace{(1-p)\cdots(1-p)}_{n-k}\cdot\underbrace{p\cdots p}_{k} = (1-p)^{n-k}p^k$$

which is the same probability. In short, we will always get the same number – only the order of the factors will vary. Thus, all we have to do is determine how many different orders are possible, that is, how many different n-tuples there are with exactly k s's. But we know that this number is $C(n,k)$. Hence, we must add up the number $p^k(1-p)^{n-k}$ a total of $C(n,k)$ times, and this gives the desired probability

$$C(n,k)p^k(1-p)^{n-k}$$

★

Example 14

Six fair coins are flipped. What is the probability of getting exactly 3 heads?

Solution
This is a binomial experiment with $n = 6$, $k = 3$ and the probability of success (getting a heads) is $p = \frac{1}{2}$. Hence, according to Theorem 2,

$$\begin{aligned}\mathcal{P}(\text{getting exactly 3 heads}) &= C(6,3)\left(\frac{1}{2}\right)^3\left(\frac{1}{2}\right)^3 \\ &= \frac{C(6,3)}{2^6} = \frac{20}{64} = \frac{5}{16} = 0.3125\end{aligned}$$ ★

Example 15

Four cards are drawn, with replacement, from a deck of cards. What is the probability of getting at least 3 aces?

Solution
The probability of getting *at least* 3 aces is equal to the probability of getting exactly 3 aces *plus* the probability of getting exactly 4 aces. Since we are dealing with a binomial experiment, with probability of success (getting an ace) equal to $p = \frac{4}{52} = \frac{1}{13}$, Theorem 2 gives

$$\begin{aligned}&\mathcal{P}(\text{getting at least 3 aces}) \\ &\quad = \mathcal{P}(\text{getting exactly 3 aces}) + \mathcal{P}(\text{getting exactly 4 aces}) \\ &\quad = C(4,3)\left(\frac{1}{13}\right)^3\left(\frac{12}{13}\right)^1 + C(4,4)\left(\frac{1}{13}\right)^4\left(\frac{12}{13}\right)^0 \\ &\quad = 4 \cdot \frac{12}{13 \cdot 13 \cdot 13 \cdot 13} + 1 \cdot \frac{1}{13 \cdot 13 \cdot 13 \cdot 13} \\ &\quad = \frac{4 \cdot 12 + 1}{13 \cdot 13 \cdot 13 \cdot 13} = \frac{49}{28561} \approx 0.0017\end{aligned}$$

which is quite small indeed. ★

EXERCISES

Define or discuss each of the following terms.

a) Mutually exclusive events
b) Independent events
c) Bernoulli experiment
d) Bernoulli trial
e) Success
f) Failure
g) Binomial experiment

The Complement of an Event

1. Two cards are drawn at random from a deck of 52 cards, but are not replaced after each draw. What is the probability of drawing at least one spade?

2. Two cards are drawn at random from a deck of 52 cards, each one being replaced after it is drawn. What is the probability of getting at least one ace?
3. Repeat the previous exercise assuming that the cards are not replaced after each drawing.
4. A fair coin is tossed 5 times. What is the probability of getting at most 4 heads?
5. A certain test consists of 10 multiple choice questions. Each question has four choices. What is the probability of getting at most 9 questions correct?
6. A pair of fair dice is rolled. What is the probability of getting two *different numbers* on the two dice?
7. A pair of fair dice is rolled. What is the probability of getting a sum that is not equal to 7?
8. Three cards are drawn at random from a deck of cards, each one being replaced after it is drawn. What is the probability that at least two of the three cards are different?

The Union and Intersection of Events

Suppose you draw a card from a deck of cards. Let E be the event of getting a diamond, let F be the event of getting a face card, let H be the event of getting a card with face value at least 9. (This includes face cards.) Describe the following events, both in words and in symbols.

9. $E \cup H$
10. $E \cap H$
11. $F \cup H$
12. $F \cap H$
13. $(E \cup F) \cap H$
14. $E \cup (F \cap H)$

Suppose that you roll two fair dice. Let E be the event of getting a sum of 7, let F be the event of getting an even sum, and let H be the event that the number on at least one of the dice is odd. Describe the following events, both in words and in symbols.

15. $E \cap F$
16. $E \cap H$
17. $F \cap H$
18. $E \cup F$
19. $E \cup H$
20. $F \cup H$

21. A card is drawn at random from a deck of cards. What is the probability that it will be a heart or a face card?
22. Suppose you toss a fair coin twice. Let E be the event of getting a head on the first toss, and let F be the event of getting a tail on the second toss.
    a) Are these events mutually exclusive? Explain your answer.
    b) What is the probability of getting a head on the first toss or a tail on the second toss? Answer this question using Theorem 1 and then using the sample space itself.
23. Suppose you toss three fair coins – one penny, one nickel and one dime.
    a) What is the probability that the penny will be a heads *and* that the nickel and dime will be the same (both heads or both tails)?
    b) What is the probability that the penny will be a heads *or* that the nickel and dime will be the same (both heads or both tails)?
24. A fair die is rolled and a fair coin is tossed. What is the probability that either the die is a 1 and the coin is a heads or the die is a 2 and the coin is a tails?

25. A fair die is rolled and a card is drawn at random.
    a) What is the probability that the die is even *and* the card is a club?
    b) What is the probability that the die is even *or* the card is a club?
26. A card is drawn at random from a deck of cards. What is the probability that it will be a club or be an even numbered card?
27. Two fair dice are rolled. What is the probability that the dice are both the same or that the sum is 6?
28. Two fair dice are rolled. What is the probability that the first die is even or that the sum is odd?

Independent Events

In Exercises 29–32, decide whether or not the events described are independent. Give reasons for your decision.

29. Experiment: rolling a die and flipping a coin. Events: getting a 4 on the die, and getting a head on the coin.
30. Experiment: drawing two cards from a deck of cards, without replacing the first card. Events: getting a spade on the first card, and getting a heart on the second card.
31. Experiment: drawing two cards from a deck of cards, but replacing the first card before drawing the second. Events: getting an ace on the first card, and getting a heart on the second card.
32. Experiment: tossing the same coin twice. Events: getting a tail on the first toss, and getting a tail on the second toss.
33. Suppose you divide an ordinary deck of cards into two decks. Deck 1 consists of all of the red cards, and deck 2 consists of all of the black cards. Consider the experiment of first flipping a coin, and then choosing a card from deck 1 if the coin shows heads, and choosing a card from deck 2 if the coin shows tails. Are the events of getting a heads and getting an ace independent?
34. Suppose you divide an ordinary deck of cards into two decks. Deck 1 consists of all of the red cards and all face cards (jacks, queens and kings), and deck 2 consists of the remaining cards. Consider the experiment of first flipping a coin, and then choosing a card from deck 1 if the coin shows heads, and choosing a card from deck 2 if the coin shows tails. Are the events of getting a heads and getting a five independent?
35. Consider the experiment of rolling a fair die and flipping a fair coin. Use the concept of independence to compute the probability that you will get a 3 on the die and a tails on the coin.
36. Suppose that you roll a fair die 3 times, and then flip a fair coin 5 times. Use the concept of independence to compute the probability that you will get the sequence

$$1 - 2 - 1 - H - H - T - H - T$$

37. Suppose you draw 4 cards from a deck of cards, replacing each card before choosing the next.
    a) What is the probability that you will draw 4 aces in the order $A\spadesuit - A\heartsuit - A\diamondsuit - A\clubsuit$?
    b) What is the probability that you will draw 4, not necessarily distinct, aces in any order?

38. Suppose you draw 4 cards from a deck of cards, but do not replace the cards before choosing the next.
    a) What is the probability that you will draw 4 aces in the order $A♠ - A♥ - A♦ - A♣$?
    b) What is the probability that you will draw 4 aces in any order?
39. Referring to the game of shuffleboard described in Example 10, if two pucks are pushed, one at a time, what is the probability that at least one puck lands in region 1?

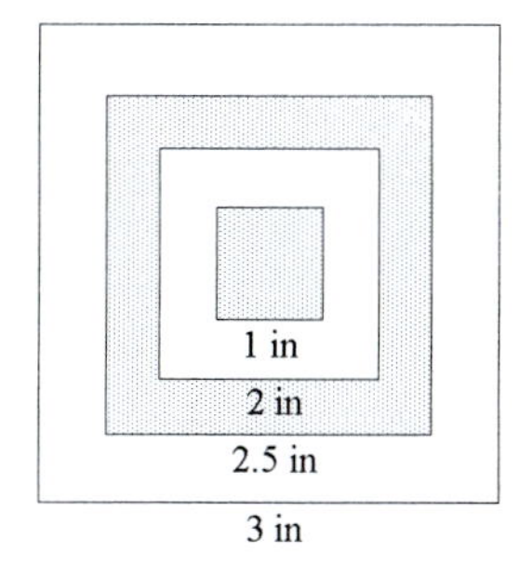

40. Consider the square dart board shown above. Imagine that two darts are thrown at the board, each one equally likely to land anywhere on the board. Assume also that the darts do not interfere with each other. What is the probability that at least one dart lands in the center square?

Binomial Probability

Suppose that you flip a fair coin 4 times. What is the probability of getting

41. exactly 1 tail?
42. exactly 2 tails?
43. exactly 3 tails?
44. no tails?
45. at least 3 heads?
46. at least 2 heads?
47. at least 1 head? (*Hint:* compute the probability of getting no heads!)

Suppose that you roll a fair die 5 times. What is the probability of getting

48. no 1's?
49. exactly one 1?
50. exactly two 1's?
51. exactly three 1's?
52. exactly four 1's?
53. exactly five 1's
54. at least four 1's?
55. at least three 1's?
56. at least one 1?

Suppose you draw a card from a deck of cards 4 times, with replacement. What is the probability of getting

57. exactly 1 ace?
58. exactly 2 aces?
59. at least 2 aces?
60. exactly 3 red cards?
61. exactly 2 clubs?
62. at least 2 clubs?
63. at least 3 clubs?
64. at least 1 face card?

65. Imagine a die that was manufactured incorrectly, and has two faces marked with a 1 and no face with a 6. Assume that each *face* is equally likely to occur. If this die is rolled 3 times, what is the probability of getting exactly two 1's?
66. In a certain pair of dice, one colored red and the other colored blue, there was a mistake in manufacturing the red die, which has two faces marked with a 1 and no face with a 2. If the pair of dice are rolled, what is the probability that the sum is equal to 2? Assume that each *face* is equally likely to occur on each die. This problem can be viewed as an equal-probability problem or as an unequal–probability problem, depending on the sample space you use. Explain this.
67. In your sock drawer, you have 4 pairs of socks. Each pair of socks is a different color, and the individual socks are loose in the drawer. Suppose that you reach in the drawer without looking, and remove 3 individual socks.
    a) Is this a binomial experiment? Why?
    b) What is the probability of getting two socks of the same color? (*Hint:* label the socks A1, A2, B1, B2, C1, C2 and D1, D2, where A1 and A2 have the same color, and so on. Use the fundamental counting principle to determine how many 3-tuples of socks have two socks the same color. Choose that color first, then choose the third sock, and then determine how many ways your choices can be distributed into a 3-tuple.)

A random walk

A drunken man, pictured in the figure below, can't make up his mind which way to walk. He decides to flip a fair coin 8 times. Each time the coin comes up heads, he walks one block west, towards the police station. Each time the coin comes up tails, he walks one block east, towards the hospital. (The drunk can walk past either the hospital or the police station.) This is referred to as a ***random walk***.

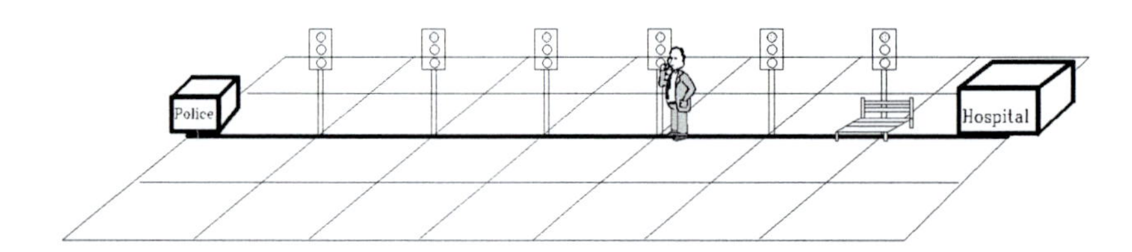

68 What is the probability that the man ends up in front of the police station?
69. What is the probability that the man ends up in front of the hospital?
70. What is the probability that the man ends up in front of the park bench?
71. What does the formula given in Theorem 2 look like when the probability of success is $\frac{1}{2}$? Can you relate this to a discussion we had near the end of Section 2.1?

2.3 Conditional Probability; Expected Value

Conditional Probability

As you know, in order to compute the probability of an event, we must first determine the possible outcomes, that is, we must first determine the sample space. It often happens, however, that we may subsequently obtain additional information about the possible outcomes, and this can have an effect on the sample space, thereby changing the probabilities. Let us illustrate with a typical example.

Example 1

Suppose that a family with 2 children has just moved into your neighborhood, and you have not yet seen them. With no additional information about the children, you would say that the sample space in this case is

$$S_1 = \{(B,B),(B,G),(G,B),(G,G)\}$$

where, for instance, (B,G) represents the possibility that the older child is a boy and the younger child is a girl. We shall assume that each child is equally likely to be a boy or a girl and therefore each of the outcomes in S_1 is equally likely to occur. The event of having two boys is $E = \{(B,B)\}$, and so the probability of two boys is

$$\mathcal{P}(\text{two boys}) = \frac{|E|}{|S_1|} = \frac{1}{4}$$

Now suppose you meet one of the parents, who tells you that their older child is a boy. This changes the sample space, since the outcomes (G,B) and (G,G) are no longer possible. In fact, the sample space is now

$$S_2 = \{(B,B),(B,G)\}$$

and we get

$$\mathcal{P}(\text{two boys, knowing that the older is a boy}) = \frac{|E|}{|S_2|} = \frac{1}{2} \qquad (1)$$

Finally, let us assume that the parents have told you that *one* of their children is a boy, but they didn't tell you which one (the older or the younger). Since this rules out the possibility (G,G), the sample space is now

$$S_3 = \{(B,B),(B,G),(G,B)\}$$

and the probability becomes

$$\mathcal{P}(\text{two boys, knowing that one is a boy}) = \frac{|E|}{|S_3|} = \frac{1}{3} \tag{2}$$

The probabilities that we computed in the previous example are referred to as **conditional probabilities**. In particular, the probability in equation (1) is the *conditional probability of two boys, given that the older is a boy*, and the probability in equation (2) is *the conditional probability of two boys, given that at least one is a boy*. It is customary to use the notation $\mathcal{P}(E \mid F)$ to denote the conditional probability of E given F.

Fortunately, there is a simple formula that can be used to compute conditional probabilities. In fact, we use the formula as our definition.

Definition

Let S be a sample space and suppose that E and F are events. Then the conditional probability of E, given that F has occurred, is

$$\mathcal{P}(E \mid F) = \frac{\mathcal{P}(E \cap F)}{\mathcal{P}(F)} \tag{3}$$

Notice that this formula involves the probability of the intersection of the two events E and F. In computing conditional probabilities, we may always use formula (3), since this is the definition. However, in the case where all outcomes are equally likely, as in Example 1, we may instead choose to determine the new sample space, based on the given information, and then compute the conditional probability directly from this new sample space.

Before doing another example of computing conditional probabilities, let us see why these two approaches give the same result in the case of equally probable outcomes. By examining Example 1, you can see that it is the given event F that becomes the new sample space when computing the conditional probability $\mathcal{P}(E \mid F)$. Also, the outcomes we are interested in are those that belong to the original event E *and* to the new sample space F. In other words, the outcomes we are interested in belong to the intersection $E \cap F$. Hence, in the case of equally probable events, we have

$$\mathcal{P}(E \mid F) = \frac{|E \cap F|}{|F|}$$

If we divide the numerator and denominator by the size $|S|$ of the *original* sample space, the result is

$$\mathcal{P}(E \mid F) = \frac{|E \cap F|}{|F|} = \frac{|E \cap F|/|S|}{|F|/|S|} = \frac{\mathcal{P}(E \cap F)}{\mathcal{P}(F)}$$

which is just formula (3).

Example 2

Suppose you roll 2 fair dice.

a) What is the probability that the two numbers on the top faces of the dice are the same?
b) What is the conditional probability that the two numbers are the same, given that the sum of these numbers is 6?
c) What is the conditional probability that the two numbers are the same, given that the sum of these numbers is 7?

Solution

a) With no additional knowledge, we do not have a conditional probability. The sample space consists of the 36 ordered pairs

$$S = \{(1,1), (1,2), \ldots, (6,6)\}$$

each of which are equally likely. The event E of getting two identical numbers is

$$E = \{(1,1), (2,2), (3,3), (4,4), (5,5), (6,6)\}$$

and since $|E| = 6$, we have

$$\mathcal{P}(\text{two identical numbers}) = \frac{6}{36} = \frac{1}{6}$$

b) Here we have a conditional probability, since we are given the additional fact that the sum of the numbers on the dice is 6.

Approach 1: The first approach is to look through the original sample space, and pick out only those outcomes whose sum is 6. This gives the new sample space

$$F = \{(1,5), (2,4), (3,3), (4,2), (5,1)\}$$

which is just the event of getting a sum of 6. Now, the event H of getting two identical numbers is

$$H = \{(3,3)\}$$

and so

$$\mathcal{P}(\text{two identical numbers} \mid \text{sum is } 6) = \frac{|H|}{|F|} = \frac{1}{5}$$

Approach 2: The second approach is to use formula (3). For this, we let F be the event of getting a sum of 6. Since there are 5 outcomes that give a sum of 6, we have

$$\mathcal{P}(F) = \frac{5}{36}$$

Let E be the event that the two numbers are the same. Then $E \cap F$ is the event of getting a sum of 6 and having both numbers the same, which happens only when the outcome is $(3,3)$. Hence

$$\mathcal{P}(E \cap F) = \frac{1}{36}$$

We can now use formula (3) to get

$$\mathcal{P}(E \mid F) = \frac{\mathcal{P}(E \cap F)}{\mathcal{P}(F)} = \frac{1/36}{5/36} = \frac{1}{5}$$

just as with the first approach.

c) Again we have a conditional probability.

Approach 1: Since the sum of the numbers must be 7, the new sample space is

$$F = \{(1,6),(2,5),(3,4),(4,3),(5,2),(6,1)\}$$

and since none of these outcomes has two identical numbers, the event H of getting two identical numbers is the empty set $\emptyset$! Thus,

$$\mathcal{P}(\text{two identical numbers} \mid \text{sum is } 7) = \frac{|H|}{|F|} = \frac{0}{6} = 0$$

In other words, if the sum is 7, then the two numbers on the top faces of the dice cannot be equal.

Approach 2: Let F be the event of getting a sum of 7. Since there are 6 outcomes whose sum is 7, we have

$$\mathcal{P}(F) = \frac{6}{36}$$

Let E be the event of having both numbers the same. Then $E \cap F$ is the event of getting a sum of 7 with both numbers the same. But this cannot happen and so $E \cap F = \emptyset$. Thus,

$$\mathcal{P}(E \cap F) = 0$$

and formula (3) gives

$$\mathcal{P}(E \mid F) = \frac{\mathcal{P}(E \cap F)}{\mathcal{P}(F)} = \frac{0}{6/36} = 0$$

just as with the first approach.★

Example 3

A certain operation results in complete recovery 60% of the time, partial recovery 30% of the time and death 10% of the time. What is the probability of complete recovery, given that a patient survives the operation?

Solution Let E be the event of complete recovery and let F be the event of surviving the operation. Then F^c is the event of dying and so

$$\mathcal{P}(F) = 1 - \mathcal{P}(F^c) = 1 - \frac{1}{10} = \frac{9}{10}$$

The intersection $E \cap F$ is the event that the patient recovers completely and survives, which is just E. Hence

$$\mathcal{P}(E \cap F) = \mathcal{P}(E) = \frac{6}{10}$$

and so

$$\mathcal{P}(E \mid F) = \frac{\mathcal{P}(E \cap F)}{\mathcal{P}(F)} = \frac{\mathcal{P}(E)}{\mathcal{P}(F)} = \frac{6/10}{9/10} = \frac{2}{3} \approx 0.67 \qquad \bigstar$$

Sometimes we are given the conditional probability $\mathcal{P}(E \mid F)$ and the probability $\mathcal{P}(F)$ of F and we wish to find the probability $\mathcal{P}(E \cap F)$ that both events occur. Formula (3) is useful in these situations as well, when written in the form

$$\mathcal{P}(E \cap F) = \mathcal{P}(E \mid F) \cdot \mathcal{P}(F) \qquad (4)$$

Example 4

A company that manufactures computer chips uses two different manufacturing processes. Process 1 produces nondefective chips 98.5% of the time and process 2 produces nondefective chips 97.1% of the time. Process 1 is used 60% of the time. What is the probability that a randomly chosen chip was produced by process 2 and is nondefective?

Solution Let E be the event that the chip is nondefective and let F be the event that the chip was produced by process 2. We seek the probability $\mathcal{P}(E \cap F)$ that both events have occurred. Since process 2 is used 40% of the time, the probability that a given chip was made by process 2 is

$$\mathcal{P}(F) = 0.40$$

We are also given that the conditional probability that the chip is nondefective, given that it was produced by process 2, is

$$\mathcal{P}(E \mid F) = 0.971$$

(Note that $97.1\% = 0.971$.) Hence, according to formula (4),

$$\mathcal{P}(E \cap F) = \mathcal{P}(E \mid F) \cdot \mathcal{P}(F) = (0.971)(0.40) = 0.3884$$

We can also write this more suggestively as

$$\begin{aligned}\mathcal{P}(\text{nondefective and process 2}) & \qquad \bigstar \\ &= \mathcal{P}(\text{nondefective} \mid \text{process 2}) \cdot \mathcal{P}(\text{process 2}) \\ &= (0.971)(0.40) = 0.3884\end{aligned}$$

An Application of Conditional Probability

One of the nicest features of formula (3) is that it can be used in a variety of interesting other formulas. Here is one example. Often times, we are interested in experiments that have two *stages*, where what we do in the second stage depends on the outcome of the first stage. Let us illustrate this.

Example 5

Imagine the following experiment. You have an unfair coin, whose probabilities are

$$\mathcal{P}(\text{heads}) = \frac{2}{3}, \mathcal{P}(\text{tails}) = \frac{1}{3}$$

You also have two urns containing colored balls, where

urn 1 has 3 blue balls and 5 red balls

urn 2 has 7 blue balls and 6 red balls

First you toss the coin. If the coin comes up heads, you draw a ball at random from urn 1. If the coin comes up tails, you draw a ball at random from urn 2. What is the probability that the ball drawn is blue? ★

The experiment in Example 5 has two stages. The first stage is the flipping of the coin, and the second stage is the drawing of a ball. Note that what you do during the second stage depends on the outcome of the first stage.

To solve problems such as the one posed in Example 5, we have a formula, which we put into a theorem.

Theorem 1

Let E and F be events from the same experiment. Let F^c be the complement of the event F. Then

$$\mathcal{P}(E) = \mathcal{P}(E \mid F)\mathcal{P}(F) + \mathcal{P}(E \mid F^c)\mathcal{P}(F^c) \qquad (5)★$$

In words, formula (5) says

$$\begin{aligned}&\text{Probability of } E\\ &\quad = (\text{Probability of } E \text{ given } F) \times (\text{Probability of } F)\\ &\quad\quad + (\text{Probability of } E \text{ given } F^c) \times (\text{Probability of } F^c)\end{aligned}$$

Let us take a moment to describe the origins of formula (5). Suppose that E and F are two events in a sample space S. Clearly, every element of E is either in F or not in F, that is, every element of E is either in F or in F^c. The portion of E that lies in F is the intersection $E \cap F$ and the portion of E that lies in F^c is $E \cap F^c$. Hence, we have

$$E = (E \cap F) \cup (E \cap F^c)$$

The events $E \cap F$ and $E \cap F^c$ are mutually exclusive, since no outcome can be in both $E \cap F$ and $E \cap F^c$. We saw in the previous chapter that when two events are mutually exclusive, the probability of their union is just the sum of their probabilities and so

$$\mathcal{P}(E) = \mathcal{P}(E \cap F) + \mathcal{P}(E \cap F^c) \qquad (6)$$

Now we use formula (3) for conditional probability, which we can write in the form

$$\mathcal{P}(E \cap F) = \mathcal{P}(E \mid F)\mathcal{P}(F)$$

and similarly,

$$\mathcal{P}(E \cap F^c) = \mathcal{P}(E \mid F^c)\mathcal{P}(F^c)$$

Substituting these two formulas into formula (6) gives formula (5).

Now let us see how formula (3) works on our example.

EXAMPLE 5 continued

According to the problem, which urn we use at stage 2 of the experiment depends on the outcome of the coin toss. If the coin comes up heads, we use urn 1, in which case the probability of drawing a blue ball is $\frac{3}{8}$, since there are 3 blue balls out of 8 balls in urn 1. In other words, the conditional probability of drawing a blue ball, given that the coin is heads, is $\frac{3}{8}$. In symbols

$$\mathcal{P}(\text{blue ball} \mid \text{heads}) = \frac{3}{8}$$

Similarly,

$$\mathcal{P}(\text{red ball} \mid \text{heads}) = \frac{5}{8}$$

On the other hand, if the coin comes up tails, we use urn 2, which has 7 blue and 6 red balls. Thus, the conditional probabilities are

$$\mathcal{P}(\text{blue ball} \mid \text{tails}) = \frac{7}{13}$$

and

$$\mathcal{P}(\text{red ball} \mid \text{tails}) = \frac{6}{13}$$

To use formula (5), we let F be the event of getting a heads on the coin, and then F^c is the event of getting a tails. The probability of each of these events is given to us in the problem, and formula (5) gives

$$\begin{aligned}
&\mathcal{P}(\text{blue ball}) \\
&\quad = \mathcal{P}(\text{blue ball} \mid \text{heads})\mathcal{P}(\text{heads}) + \mathcal{P}(\text{blue ball} \mid \text{tails})\mathcal{P}(\text{tails}) \\
&\quad = \frac{3}{8} \cdot \frac{2}{3} + \frac{7}{13} \cdot \frac{1}{3} = \frac{1}{4} + \frac{7}{39} = \frac{67}{156} \approx 0.4295
\end{aligned}$$

★

Expected Value

Now we turn our attention to a slightly different matter. Suppose that someone has offered to play a certain game of chance with you. The rules are as follows.

A single fair die is rolled. If the outcome (number on top face) is a 1, 2, 3 or 4, you win the number of dollars shown on the die. If the outcome is a 5 or 6, you lose that many dollars. Would you play this game?

Since the die is fair, the probability of getting any particular number is $\frac{1}{6}$. Thus, for instance, the probability is $\frac{1}{6}$ that a 3 will appear, and so you will win 3 dollars with probability $\frac{1}{6}$. In other words, $\frac{1}{6}$-th of the time, on the average, you will win 3 dollars. For convenience, it is easier to think of losing as winning a negative amount, and so we can say

you win 1 dollar with probability $\frac{1}{6}$
you win 2 dollars with probability $\frac{1}{6}$
you win 3 dollars with probability $\frac{1}{6}$
you win 4 dollars with probability $\frac{1}{6}$
you win -5 dollars with probability $\frac{1}{6}$
you win -6 dollars with probability $\frac{1}{6}$

Thus, you can expect, *on the average*, to win a total of

$$\begin{aligned} 1\cdot\frac{1}{6}+2\cdot\frac{1}{6}+3\cdot\frac{1}{6}+4\cdot\frac{1}{6}+(-5)\cdot\frac{1}{6}+(-6)\cdot\frac{1}{6} & \\ &= \Big(1+2+3+4+(-5)+(-6)\Big)\cdot\frac{1}{6} \\ &= (-1)\cdot\frac{1}{6} \\ &= -\frac{1}{6}\text{ dollars} \end{aligned} \tag{7}$$

That is, you can expect to *lose* $\frac{1}{6}$-th of a dollar (about 17 cents), on the average, by playing this game. Thus, it would not be wise to play this game.

We should emphasize that you will never actually lose $\frac{1}{6}$-th of a dollar in any given play. The computation above tells you what will happen, on the average, over a long period of time. For instance, if you were to play this game 1000 times, you would expect to lose approximately

$$\frac{1000}{6} = 167\text{ dollars}$$

The number computed in equation (7) is called your *expected value* for this game. Let us make a formal definition of this important concept.

Definition

Suppose that a game has the possible outcomes $s_1, s_2, \ldots, s_n$, and that you win the amount w_i (which may be negative) when outcome s_i occurs, for all $i = 1, 2, \ldots, n$. Then your **expected value** for this game is

$$\mathcal{E} = w_1\cdot\mathcal{P}(s_1) + w_2\cdot\mathcal{P}(s_2) + \cdots + w_n\cdot\mathcal{P}(s_n)$$ ★

Unfortunately, the term *expected value* is a bit misleading since, as we said before, the expected value is not necessarily a number that you expect to win, but rather it is an indication of what should happen, on the average, over a long period of time. When the expected value is positive, then in the long run you expect to have positive winnings, in proportion to the expected value. If the expected value is negative, then you expect in the long run to have negative winnings, that is, you expect to lose.

Example 6

A radio station is having a contest. The first prize is a car worth \$10,000. There are 2 second prizes, each of which is a stereo worth \$1,000, and there are 5 third prizes, each of which is a 1-year supply of puppy food, worth \$150 (assuming that you have a puppy). There is no entry fee for this contest, but you must send in a post card, which costs 15 cents. Assuming that 100000 people enter this contest, should you enter?

Solution

To answer this question, we must compute your expected value for this contest. We begin with the various probabilities

$$\mathcal{P}(\text{winning first prize}) = \tfrac{1}{100000}$$

$$\mathcal{P}(\text{winning second prize}) = \tfrac{2}{100000}$$

$$\mathcal{P}(\text{winning third prize}) = \tfrac{5}{100000}$$

$$\mathcal{P}(\text{losing}) = \tfrac{99992}{100000}$$

Now, remembering that even when we win a prize, we still lose the 15 cents for the post card, we get (with the help of a calculator) the expected value

$$\begin{aligned}
\mathcal{E} &= (10000 - 0.15) \cdot \frac{1}{100000} + (1000 - 0.15) \cdot \frac{2}{100000} \\
&\quad + (150 - 0.15) \cdot \frac{5}{100000} + (-0.15) \cdot \frac{99992}{100000} \\
&= \frac{1}{100000}\Big(9999.85 + 2 \cdot 999.85 + 5 \cdot 149.85 - (0.15 \cdot 99992)\Big) \\
&= -\frac{2250}{100000} \approx -0.0225
\end{aligned}$$

Thus, you can expect, on the average, to lose 2.25 cents every time you enter such a contest. From this point of view, you would not be wise to enter.

However, we should mention that there may be other factors involved in your decision to enter the contest, besides the mathematical expectation of winning. For instance, you may consider that losing 15 cents will not make any real difference in your life, in comparison with winning a new car. From this point of view, you may reasonably choose to enter the contest. In summary, the expected value is simply a mathematical expression of what you can expect to happen over the long run, and it should not be taken as anything more.

Nevertheless, in more complicated situations, it may be the best (or only) guide as to what to do. ★

EXERCISES

Define or discuss each of the following terms.

a) Conditional probability   b) Expected value

Conditional probability

A fair die is rolled. Compute the following probabilities.

1. The probability of getting a 5.
2. The conditional probability of getting a 5, given that the outcome is odd.
3. The conditional probability of getting a 5, given that the outcome is at least 4.

A pair of fair dice are rolled. One of the dice is red, the other is white. Compute the following probabilities.

4. The probability of getting a sum of 5.
5. The conditional probability of getting a sum of 5, given that the red die is a 3.
6. The conditional probability of getting a sum of 5, given that one of the dice is a 3.
7. The conditional probability of getting a sum of 5, given that the white die is even.
8. The conditional probability of getting a sum of 5, given that one of the dice is even.
9. The conditional probability of getting a sum of 6, given that the sum is even.
10. The probability that the two dice are the same.
11. The conditional probability that the two dice are the same, given that the red die is a 4.
12. The conditional probability that the two dice are the same, given that one die is a 4.

A single card is drawn from a deck of cards. Compute the following probabilities.

13. The probability of getting a spade.
14. The conditional probability of getting a spade, given that the card is black.
15. The conditional probability of getting a spade, given that the card is a queen.
16. The conditional probability of getting a face card, given that the card is black.

Two cards are drawn from a deck of cards, one-at-a-time, without replacement. Compute the following probabilities.

17. The conditional probability that the second card is a king, given that the first card is an ace.
18. The conditional probability that the second card is a king, given that the first card is a king.
19. The conditional probability that the second card is a king, given that the first card is a heart.
20. The conditional probability that the two cards have the same denomination, given that the first card is a king.
21. The conditional probability that the two cards have the same denomination, given that at least one of the cards is a king.

Three coins are flipped. Compute the following probabilities.

22. The probability that the third coin is a head, given that the first two are heads.
23. The conditional probability that all three flips result in the same outcome, given that the first coin is a tail.
24. The conditional probability that all three flips result in the same outcome, given that at least one coin is a tail.

A card is drawn, and a single die is rolled. Compute the following probabilities.

25. The probability that the number on the card is the same as the number on the die.
26. The conditional probability that the number on the card is the same as the number on the die, given that the number on the card is a 5.
27. The conditional probability that the number on the card is the same as the number on the die, given that the number on the card is a 10.

A certain family has three children. Compute the following probabilities. Assume that the probability of having a boy is equal to the probability of having a girl, and that the gender of one child has no effect on the gender of future children.

28. The probability that all of the children are girls.
29. The conditional probability that all of the children are girls, given that the oldest one is a girl.
30. The conditional probability that all of the children are girls, given that at least one of the children is a girl.
31. The conditional probability that all of the children are girls, given that at least two of the children are girls.
32. Studies of the racing history of a certain race horse show that the horse wins 12% of its races on dry tracks, but only 6% of its races on wet tracks. A certain race track is dry 80% of racing days and wet 20% of racing days. (a) What is the probability that, on a randomly chosen day, the track is wet and the horse wins? (b) What is the probability that, on a randomly chosen day, the track is wet and the horse loses?
33. A certain rocket is to be launched. Past history shows that the probability that the launch is a "go" on a sunny day is 0.95, the probability that the launch is a "go" on an overcast day is 0.75 and the probability that the

launch is a "go" on a rainy day is 0.35. The weather bureau reports that, at the launch site, 89% of the days are sunny, 7% of the days are overcast and 4% of the days are rainy. (a) What is the probability that, on a randomly chosen day, the weather is rainy and the launch is a "go"? (b) What is the probability that, on a randomly chosen day, the weather is overcast and the launch is a "go"? (c) What is the probability that, on a randomly chosen day, the weather is sunny and the launch is a "go"?

34. Three cards lie in a box. One of the cards is white on both sides, one is black on both sides, and one is white on one side, and black on the other. Suppose you reach into the box without looking, draw a card at random, and place it on a table. If the top side of the card is black, what is the probability that the underside of the card is also black? (Hint: picking a card and placing it on the table amounts to making a choice at random from among the *six* possible sides of the cards. Hence, the initial sample space has size 6.)
35. An unfair coin is tossed. The probabilities are

$$\mathcal{P}(\text{heads}) = \frac{3}{4}, \mathcal{P}(\text{tails}) = \frac{1}{4}$$

If heads comes up, you draw a ball from an urn containing 4 green balls and 4 yellow balls. If tails comes up, you draw a ball from an urn containing 5 green balls and 10 yellow balls. What is the probability of drawing a green ball? A yellow ball?

36. A fair die is rolled. If the number is 4 or less, the die is rolled once more and the number is recorded. If the first roll comes up 5 or more, the die is rolled twice more and the sum of the two numbers is recorded.
    a) What is the probability that the number recorded is 5?
    b) What is the probability that the number recorded is 11?
37. In an upcoming congressional election, two democrats, Al and Bill, are seeking the nomination to run against the incumbent republican George. It is estimated that George has probability $\frac{1}{2}$ of beating Al in the election if Al is the democratic nominee but only a $\frac{1}{3}$ probability of beating Bill. Given that Al has a $\frac{2}{3}$ probability of being the democratic nominee, what is the probability that George will be reelected?
38. The Starship Enterprise is expecting an attack from a renegade Klingon warship, with the possible help of a Romulan ship. According to Mr. Data, the probability that the Romulans will join with the Klingons is 0.125. If they do, Mr. Data estimates that the probability that the joint force will defeat the Enterprise is 0.6275, whereas the probability that the Enterprise will be victorious if the Romulans do not interfere is 0.8125. What is the probability that the Enterprise will be victorious?
39. A light bulb manufacturer produces light bulbs by two different methods. The probability that a light bulb made by the first method is defective is 0.001. The probability that a light bulb made by the second method is defective is 0.0001. Exactly $1/4$ of the total number of light bulbs are made by the first method. What is the probability that a randomly chosen light bulb will be defective?

40. A card is drawn at random from a deck of 52 cards. If the card is not a face card, a roulette wheel is spun. (An American roulette wheel has the numbers 0, 00 and 1-36.) Whatever number appears on the wheel is the amount that you win. If the card drawn is a face card, a pair of dice are rolled and you win the sum of the numbers (in dollars) on the dice. What is the probability that you will win at most $2?

Expected Value

41. Suppose that you roll a fair die once. If the number on the top face of the die is even, you win that amount, in dollars. If it is odd, you lose that amount. What is the expected value of this game? Would you play?
42. For a cost of 1 dollar, you can roll a single fair die. If the outcome is odd, you win 2 dollars. Would you play? Why?
43. Suppose you draw a card from a deck of cards. You win the amount showing on the card if it is not a face card, and lose $10 if it is a face card. What is your expected value? Would you play this game?
44. Suppose you draw a card from a deck of cards. If the card is a club, you win $3.00, but if it is any other suit, you lose $1.00. What is your expected value? Would you play this game?
45. A certain contest costs 30 postage cents to enter. There is one first prize of $20,000, two second prizes of $5000 each, and five third prizes of $100 each. If you estimate that 100,000 people will enter this contest, what is your expected value? Would you enter?
46. Suppose you toss three fair coins. You win 32 cents if all three coins come up the same. You win 8 cents if exactly 1 head occurs, and you win 48 cents if exactly 2 heads appears. Would you pay 30 cents to play this game? If not, how much would you be willing to pay?
47. A contest has one first prize of $1000, three second prizes of $200 each, and ten third prizes of $10 each. If you expect 20,000 people to enter this contest, how much would you be willing to pay to enter?
48. Imagine that you are at a grocery store. Cans of tuna cost $1.50 each. A customer mentions to you that a nearby store has discounted tuna from $1.70 per can to $1.30 per can. It will cost you about 10 cents in gas to get to the other store. If you feel that this other person is accurate about 70% of the time, would it pay (on the average) to go to the other store?
49. Consider the dart board in Exercise 17 of Section 2.1. Suppose you pay $1.50 for one dart. You are paid $3.00 if you hit the center, $2.00 if you hit the middle ring and $1.00 if you hit the outer ring. What is your expected value? Would you play this game?
50. Consider the shuffleboard game in Example 10 of Section 2.2. Suppose you push one puck. If it lands in the region numbered r, you get $1/r$ dollars. How much would you pay to play this game?

A game of chance is said to be ***fair*** *if its expected value is zero.*

51. In the simple game of cutting cards, each person puts up the same amount of money, and cuts the cards once. The person with the higher card wins. Is this game fair? Explain.

52. An American roulette wheel has 18 red numbers, 18 black numbers and two green numbers. If you bet on red, you win an amount equal to your bet (and get your original bet back) if a red number comes up, but lose your bet otherwise. What is your expected value in this game? Is this a fair game?
53. Devise your own game of chance that you think a person without a knowledge of expected values might be willing to play, to your benefit.

Chapter 3
Applications of Probability

3.1 Games of Chance

As we remarked in the previous chapter, probability theory was first developed in connection with games of chance, and in this section, we take a look at how probability theory can help us understand a simple version of poker, as well as the California State Lottery. After reading this section, you will have a clearer understanding of what your chances are for success in these games.

Draw Poker

The rules of draw poker are quite simple. (To avoid any unnecessary complications, we will simplify them a bit more.) The dealer gives each player five cards. The first player to the left of the dealer can then bet a sum of money by placing it in the center of the table. This forms the *pot*. Each player in turn either must place the same amount of money in the pot to continue to play (this is called *calling the bet)*, or withdraw from the game (this is called *folding*). (A player can also *raise* the bet by putting in more money than is required, but we will not go into this aspect further.)

Once every player has either called or folded, each remaining player, starting with the player to the left of the dealer, can exchange any number of his cards in his hand for new ones. This is called the *draw*. Once the draw is completed, another round of betting takes place. Then each player shows his cards, and the player with the highest ranking hand wins the pot.

Our interest here will center on how to compute the probability of obtaining various types of poker hands, and so we must first describe these hands. We will list them in order of ranking, highest first.

Royal flush

The highest ranking poker hand is called a **royal flush**, and consists of the ace, king, queen, jack and ten of the *same* suit, as illustrated below

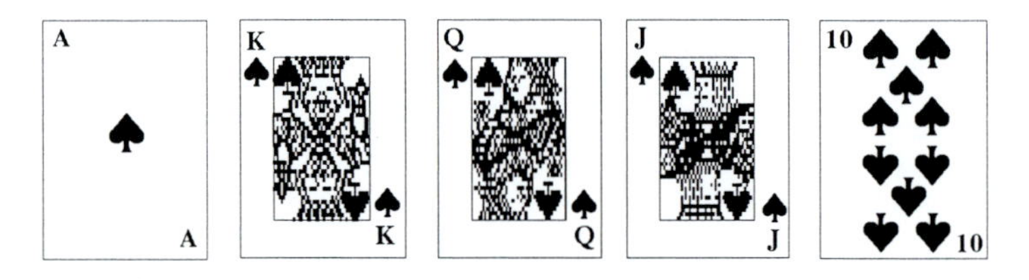

Straight flush

Any five cards in a row, and of the same suit, is called a **straight flush**.

To avoid confusion in counting, we will consider a royal flush as being *different* from a straight flush. (In other words, when we use the term straight flush, we exclude royal flushes.) To form a straight flush, an ace may be used as a low card, as in A♥2♥3♥4♥5♥.

Four of a kind

The third highest poker hand is **four of a kind**, as in the following hand, which would be described as "four sevens."

Full house

The fourth highest poker hand is the **full house**, which consists of 3 of one kind and 2 of another kind, as in

Flush

The fifth highest poker hand is the **flush**, which consists of any five cards of the same suit, as in

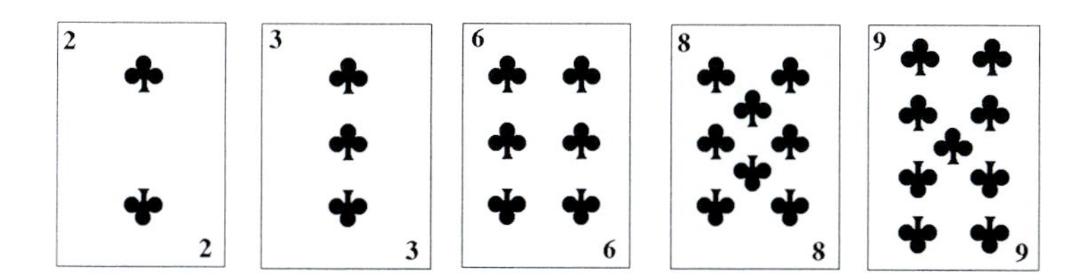

To avoid confusion, when we use the term flush, we will exclude straight flushes and royal flushes. Thus, a straight flush is *not* a flush.

Straight

Next comes the **straight**, which consists of 5 cards in a row, as in

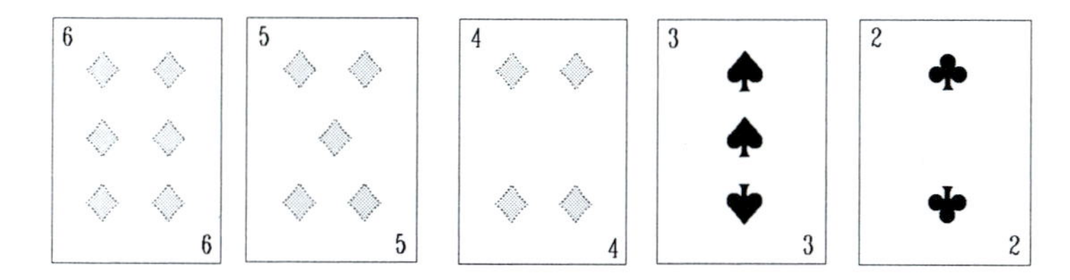

As with flushes, the term straight does not apply to straight flushes. Also, an ace can be used as either a high card, as in A♠K♦Q♦J♣10♦, or as a low card, as in A♦2♣3♣4♠5♠.

Three of a kind

The next highest hand is **three of a kind**, as in the "three aces" shown below

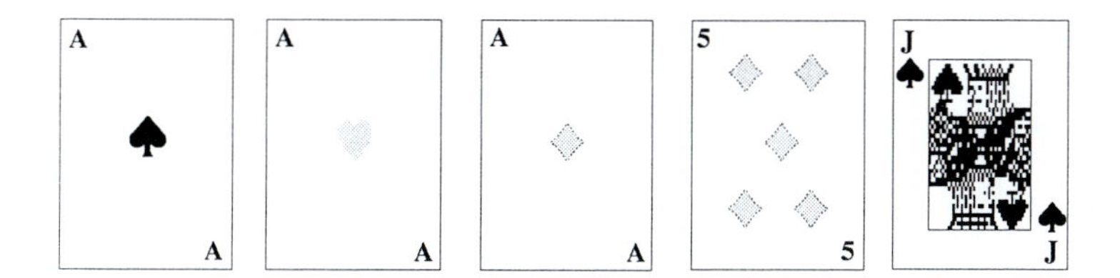

Two pairs

Next we come to **two pairs**, as in the following hand, read "eights and fives"

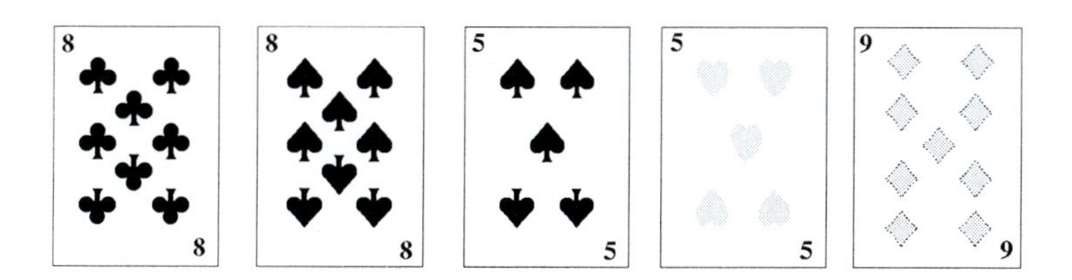

One pair
Next comes **one pair** as in

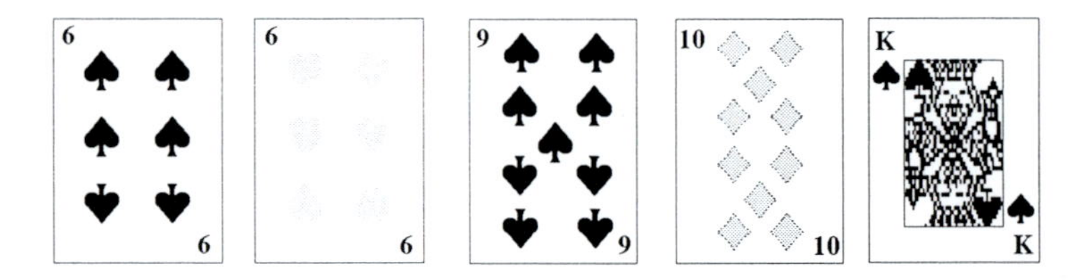

Nothing
The lowest ranking hand is **nothing**, as in

When comparing two hands that have nothing, the hand with the highest ranking card is higher than the other hand.

Now let us consider the probabilities of being dealt various poker hands. We will assume that the cards are shuffled randomly, and so getting any 5 cards is equally likely as getting any other. The sample space in this case is the set of all possible 5-card hands, and we have already seen (Section 1.3) that the sample space has size

$$C(52, 5) = 2,598,960$$

In other words, there are $2,598,960$ possible poker hands!

In order to determine the probability of being dealt a given type of poker hand, we simply count the number of such hands, and divide by $2,598,960$. Let us do a few examples, and leave the rest as exercises. Notice that we rely heavily on the fundamental counting principle.

Example 1

Determine the probability of being dealt
a) a straight flush b) a full house
c) a straight d) 3 of a kind

Solutions

a) To form a straight flush, we choose a suit, and then we choose the highest ranking card in the hand. These two tasks will completely determine the hand. For instance, if we choose the suit hearts, and the highest ranking card a 9, then the hand is 9♥8♥7♥6♥5♥. Now, there are 4 ways to choose a suit, since there are 4 suits. Also, any of the nine cards

$$5, 6, 7, 8, 9, 10, J, Q, K$$

can be chosen as the highest ranking card of the hand. We cannot choose a card lower than a 5 since we need five cards, and we cannot choose an ace, since the resulting hand would be a *royal flush*, and not a straight flush. (Remember our convention that a royal flush is not a straight flush.) Thus, there are a total of

$$4 \times 9 = 36$$

straight flushes, and so the probability of being dealt a straight flush is

$$\mathcal{P}(\text{straight flush}) = \frac{36}{2598960} \approx 0.000013851$$

which is about 1 in $72,000$! (We got this by dividing 36 into $2,598,960$, which gives $72,193$.) Thus, being dealt a straight flush is indeed a rare event.

b) In counting poker hands (and in other counting situations as well), it is often convenient to first make a list of the tasks we must perform in order to determine a particular type of hand. In this case, to determine a full house, we must

1) Choose a denomination for the 3 of a kind (for example: aces, or sevens, or threes, etc.)
2) Choose the 3 cards of that denomination
3) Choose the denomination for the 2 of a kind
4) Choose the 2 cards of that denomination

Now, there are $C(13, 1)$ ways to perform the first task, since there are 13 denominations. There are $C(4, 3)$ ways to perform the second task of choosing 3 out of the four cards of the given denomination for the hand. There are $C(12, 1)$ ways to perform the third task, since after the first task is performed, there are only 12 denominations available to choose from. Finally, there are $C(4, 2)$ ways to perform the fourth task. Hence, according to the fundamental counting principle, there are

$$C(13, 1)C(4, 3)C(12, 1)C(4, 2) = 13 \cdot 4 \cdot 12 \cdot 6 = 3744$$

full houses in the deck, and so the probability of getting a full house is

$$\mathcal{P}(\text{full house}) = \frac{3744}{2598960} \approx 0.001440576$$

which is about 1 in 700. (Divide 3744 into 2598960.)

c) A straight is determined by choosing the denomination of the highest ranking card in the hand, and then by choosing a card for each required denomination. For instance, if we choose a 9 as the highest denomination, then we must choose one 9, one 8, one 7, one 6 and one 5. Now, we can choose any of 5, 6, 7, 8, 9, 10, J, Q, K or A for the highest ranking card, and so there are 10 ways to perform this task. Also, choosing a card for each required denomination can be done in 4 ways, and so we get

$$10 \times 4 \times 4 \times 4 \times 4 \times 4 = 10 \times 4^5 = 10240$$

However, we must pause a moment, and realize that we have included straight flushes, and royal flushes, in this count. Hence, since there are 36 straight flushes, and 4 royal flushes (why?), the total number of straights is

$$10240 - 36 - 4 = 10200$$

Accordingly, the probability of getting a straight is

$$\mathcal{P}(\text{straight}) = \frac{10200}{2598960} \approx 0.003924646$$

which is about 1 in 250.

d) To determine a hand of this type, we make the following choices.
   1) Choose the denomination for the 3 of a kind
   2) Choose the 3 cards of that denomination
   3) Choose the remaining 2 cards for the hand

   The first task can be performed in $C(13,1)$ ways. The second task can be performed in $C(4,3)$ ways. For the third task, we must choose 2 cards from among the remaining $52 - 4 = 48$ cards. (We cannot choose the fourth card of the denomination selected for the 3 of a kind, since that would give us 4 of a kind.) Since this can be done in $C(48,2)$ ways, we get

$$C(13,1)C(4,3)C(48,2) = 13 \cdot 4 \cdot 1128 = 58656$$

However, we must observe that full houses have been included in this count, since the 2 cards chosen in the third task may have been the same denomination. Hence, we must subtract the number of full houses, to get

$$58656 - 3744 = 54912$$

hands consisting of 3 of a kind. Finally, the probability of being dealt 3 of a kind is

$$\mathcal{P}(3 \text{ of a kind}) = \frac{54912}{2598960} \approx 0.021128451$$

or about 1 in 50. ★

Let us consider some additional probability computations that you might want to make during a game of draw poker.

Example 2

Suppose you have been dealt 4 hearts and 1 spade. You decide to replace the spade during the draw, in the hopes of getting another heart, to make a flush. What is the probability of getting a flush?

Solution Assuming that you have not seen any other cards, as far as you know there are $52 - 5 = 47$ outstanding cards, of which $13 - 4 = 9$ are hearts. Therefore, the probability that one of these cards given to you at random on the draw is a heart is $\frac{9}{47}$. In other words, the probability that you will get your flush is $\frac{9}{47} \approx 0.191489361$.

Incidentally, if you should happen to *accidentally* see someone else's cards, feel free to use this information to alter the sample space, and thereby adjust your computation of the probabilities.★

Example 3

a) Suppose that you have been dealt $10\diamondsuit 9\clubsuit 8\spadesuit 7\diamondsuit K\clubsuit$. Your plan is to draw one card (to replace the $K\clubsuit$), in the hopes of making a straight. What is the probability of doing so?
b) Suppose that you have been dealt $10\heartsuit 9\heartsuit 7\diamondsuit 6\spadesuit A\heartsuit$. Your plan is to draw one card (to replace the $A\heartsuit$), in the hopes of making a straight. What is the probability of doing so?

Solutions

a) Without any additional information, you may as well assume that there are 47 outstanding cards, and you need any of the four Jacks, or four 6's, to make your straight. In other words, there are 8 cards that would make your straight, out of a total of 47 cards, and so the probability of making a straight is $\frac{8}{47}$. Since there are two possible denominations (Jacks or sixes) that would give you a straight, poker players refer to this situation as "drawing to an *outside* straight"

$$\mathcal{P}(\text{drawing to an outside straight}) = \frac{8}{47} \approx 0.170212766$$

b) In this case, there are 47 outstanding cards, but only one of the four 8's will make your straight. In this case, you are drawing to an *inside* straight and the probability of making the straight is $\frac{4}{47}$. Thus,

$$\mathcal{P}(\text{drawing to an inside straight}) = \frac{4}{47} \approx 0.085106382$$ ★

Example 4

What is the probability of drawing 2 cards to a flush?

Solution In this case, you have three cards of one suit, and decide to replace your other two cards, in the hopes of getting a flush. For instance, your hand might be $J\diamondsuit 8\diamondsuit 3\diamondsuit 7\clubsuit 6\heartsuit$, and you decide to draw 2 cards to replace the $7\clubsuit$ and $6\heartsuit$.

Since there are 47 outstanding cards, there are $C(47,2)$ possibilities for receiving 2 cards on the draw. Since there are $13-3=10$ outstanding diamonds, there are $C(10,2)$ pairs of diamonds that you would be thrilled to receive on the draw. Thus, the probability of getting your flush is

$$\mathcal{P}(\text{drawing 2 cards to a flush}) = \frac{C(10,2)}{C(47,2)} = \frac{45}{1081} \approx 0.041628122$$

which is about 1 in 24. (Now you can see why drawing 2 cards to a flush is a desperation play in draw poker.)★

The California State Lottery

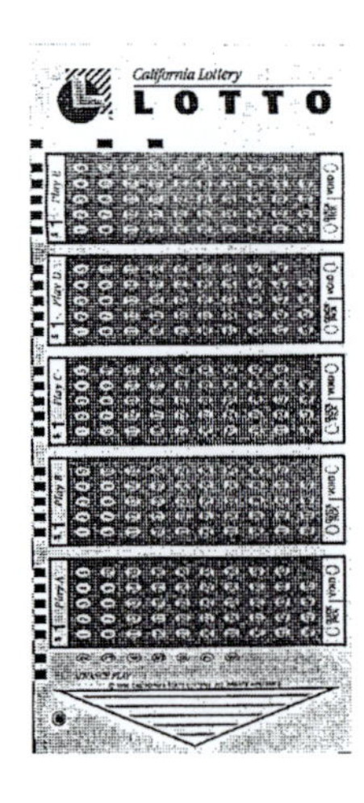

The California State Lottery, known as Lotto 6/53, is typical of state lotteries. For 1 dollar, you are entitled to choose 6 numbers from 1 to 53. Six initial numbers are drawn at random by the lottery officials, plus an additional number, referred to as the *bonus number*. The winning categories are as follows.

1) Getting all 6 initial numbers.
2) Getting 5 of the 6 initial numbers, plus the bonus number.
3) Getting 5 of the 6 initial numbers, but not the bonus.
4) Getting 4 of the 6 initial numbers.
5) Getting 3 of the 6 initial numbers.

Notice that the bonus number plays a role only if you get 5 out of 6 initial numbers. The payoffs for each winning combination are determined as follows. One half of the total ticket sales for a given drawing is set aside for expenses and for the State Lottery Education Fund. By law, a maximum of 16% of this sum can go for expenses, and so a minimum of 34% goes to the education fund. The other half of ticket sales is used for prizes.

According to lottery officials, of this half, 40% is split equally among those who are lucky enough to get all 6 initial numbers. Those who get 5 initial numbers and the bonus split 21.35% of the prize money. Those who get 5 initial numbers but not the bonus split 11% of the prize money, and those who get 4 initial numbers split 10%. The rest of the prize money (17.65%) goes into a revolving fund that is used to pay $5.00 to each person who gets 3 out of 6 initial numbers. (The fund is necessary because there may be times when more money must be paid out for category 5 winners than 17.65% of the total prize money.) You might also be interested to know that if there are no winners in categories 1-4, the prize money for those categories is all rolled over into the jackpot for *category 1*, for the next drawing.

Now, let us compute the probabilities of the various winning categories. Get your calculators warmed up.

For the sake of illustration, let us assume that the initial numbers for a given drawing are 1, 2, 3, 4, 5, and 6, and that the bonus number is 7. As you know, these numbers are just as likely to occur as any other set of numbers!

First, we need the number of possible tickets, which is

$$C(53,6) = \frac{53!}{6!47!} = 22,957,480$$

All 6 initial numbers

Since there is only one way to get all 6 initial numbers, the probability of winning the jackpot is

$$\mathcal{P}(\text{all 6 numbers}) = \frac{1}{C(53,6)} = \frac{1}{22957480} \approx 0.000000043$$

or, in other words, 1 in $22,957,480$.

Five numbers out of six, and the bonus

A ticket that has 5 initial numbers and the bonus can be formed by marking 5 out of the 6 initial numbers, and the bonus. Since there are $C(6,5)$ ways to do this, there are $C(6,5)$ possible tickets in this winning category. Hence,

$$\mathcal{P}(\text{5 out of 6 and bonus}) = \frac{C(6,5)}{C(53,6)} = \frac{6}{22957480} \approx 0.000000261$$

This is 1 in $3,826,247$. (Divide 6 into 22957480.)

Five numbers out of six, no bonus

Marking a ticket that gets 5 out of 6 initial numbers, but not the bonus, can be done in two steps. First, we choose 5 from the 6 initial numbers, which can be done in $C(6,5)$ ways. Then, we choose a sixth number that is not an initial number, nor the bonus number. This can be done in $53 - 7 = 46$ ways. Hence, the total number of tickets in this category is $46 \cdot C(6,5)$, and

$$\mathcal{P}(\text{5 out of 6, no bonus}) = \frac{46 \cdot C(6,5)}{C(53,6)} = \frac{276}{22957480} \approx 0.000012022$$

or 1 in $83,179$. (Divide 276 into 22957480.)

Four numbers out of six

To get this type of ticket, we choose 4 out of the 6 initial numbers, which can be done in $C(6,4)$ ways. Then, we choose 2 out of the $53 - 6 = 47$ noninitial numbers (we may choose the bonus), and this can be done in $C(47,2)$ ways. Hence,

$$\mathcal{P}(\text{4 out of 6}) = \frac{C(6,4) \cdot C(47,2)}{C(53,6)} = \frac{15 \cdot 1081}{22957480} \approx 0.000706305$$

or 1 in 1416.

Three numbers out of six

We will leave it as an exercise for you to show that

$$\mathcal{P}(3 \text{ out of } 6) \approx 0.014126114$$

or about 1 in 71.

Losing tickets

Tickets that have 2 or fewer initial numbers are losing tickets. We leave it as an exercise to compute the number of losing tickets, and to show that

$$\mathcal{P}(\text{losing}) \approx 0.985155255$$

and so we lose about 98.5% of the time.

To get a feel for whether or not the probabilities suggest that you should play the lotto, let us compute some expected values.

Example 5

A typical lotto sales figure is $12,000,000. Thus, assuming no rollover from the previous game, the total prize money is $6,000,000, which is distributed as discussed above

All six numbers:	40% of $6,000,000 = 2,400,000$
Five of six and bonus:	21.35% of $6,000,000 = 1,281,000$
Five of six, no bonus:	11% of $6,000,000 = 660,000$
Four of six:	10% of $6,000,000 = 600,000$

Of course, if there is more than one winner in a given category, these amounts must be divided by the number of winners. For example, a typical game might yield

Number of 6 out of 6 winners: 1
Number of 5 and the bonus winners: 5
Number of 5 and no bonus winners: 150
Number of 4 out of 6 winners: 10000

(The number of 3 out of 6 winners does not affect the amount each such winner gets, which is fixed at $5.00.) Hence, the amount going to each winner is

All six numbers: $2,400,000$
Five of six and bonus: $\frac{1281000}{5} = 256,200$
Five of six, no bonus: $\frac{660000}{150} = 4400$
Four of six: $\frac{600000}{10000} = 60$
Three of six: 5

Now we can compute the expected value, remembering that we must subtract 1 dollar from each winning total, since we do not get our dollar back. This is not significant for the larger prizes, but it is for the smaller ones.

$$\begin{aligned}\mathcal{E} &= (2400000-1)(0.000000043)+(256200-1)(0.000000261)\\&\quad+(4400-1)(0.000012022)+(60-1)(0.000706305)\\&\quad+(5-1)(0.014126114)+(-1)(0.985155255)\\&\approx -0.66402613\end{aligned}$$

and so, on the average, you expect to lose about 66 cents each time you play a lotto with these sample statistics (total prize money and number of winners.)★

EXERCISES

1. What is the probability of being dealt a royal flush in draw poker?
2. What is the probability of being dealt 4 of a kind in draw poker?
3. What is the probability of being dealt a flush in draw poker?
4. What is the probability of being dealt 2 pairs in draw poker?
5. What is the probability of being dealt 1 pair in draw poker?
6. What is the probability of being dealt nothing in draw poker?
7. What is the most probable hand in draw poker?
8. Given that you hold 8♦7♦6♣5♥2♠, what is the probability of drawing to a straight?
9. Given that you hold 9♦7♦6♣5♥2♠, what is the probability of drawing to a straight?
10. Assuming that you hold 3♦4♦5♦6♦8♣, what is the probability of drawing 1 card to a straight flush?
11. Assuming that you hold 2♦4♦5♦6♦8♣, what is the probability of drawing 1 card to a straight flush?
12. What is the probability of drawing 3 cards to a flush?
13. Assuming that you hold 6♦6♥6♠K♣A♣, what is the probability of drawing 2 cards and getting 4 of a kind?
14. Assuming that you hold 6♦6♥6♠K♣A♣, what is the probability of drawing 2 cards and getting a full house?
15. Assuming that you hold 7♠8♠9♠10♠4♥, what is the probability that, by replacing the 4♥ on the draw, you can improve your hand to either a straight, a flush or a straight flush?
16. Assuming that you hold no pairs, and draw a single card, what is the probability that you will still have no pairs?
17. Compute the probability of getting exactly 3 out of 6 initial numbers in the California State Lotto.
18. Compute the probability of getting exactly 2 out of 6 initial numbers in the California State Lotto.
19. Compute the probability of getting exactly 1 out of 6 initial numbers in the California State Lotto.
20. Compute the probability of getting none of the 6 initial numbers in the California State Lotto.
21. Compute the probability of losing in the California State Lotto.
22. Compute the expected value of the lotto assuming that total revenue is $\$12,000,000$, and that

    Number of 6 out of 6 winners: 2

Number of 5 and the bonus winners: 10
Number of 5 and no bonus winners: 250
Number of 4 out of 6 winners: 15000

23. Compute the expected value of the lotto assuming that total revenue is $\$12,000,000$, and that
    Number of 6 out of 6 winners: 1
    Number of 5 and the bonus winners: 1
    Number of 5 and no bonus winners: 100
    Number of 4 out of 6 winners: 8000
24. The California State Lotto used to be 6/49. That is, there used to be only 49 numbers to choose from, instead of 53. What is the probability of winning the grand prize in Lotto 6/49?
25. What is the expected value of the Lotto 6/49, assuming the same statistics as in Example 5?

3.2 Single Gene Inheritance

Elementary probability theory plays a fundamental role in understanding the process of genetic inheritance. In order to discuss this, however, we must begin with some terminology.

Johann Gregor Mendel
(1822–1884)
The Austrian Monk Gregor Mendel was the first to formulate the principles of heredity.

The fundamental unit of heredity is called the **gene**. Genes are transmitted from one generation to the next during reproduction, and carry coded information that allows cells to make proteins. A gene may take several different forms, however, and these forms are known as **alleles**. As a simple example, most red–green color blindness occurs genetically. The gene for color blindness has two alleles – the normal form A, indicating normal color vision, and the abnormal form a, indicating color blindness.

Human cells, called **zygotes**, have two copies of the color vision gene, but these copies may be different alleles. Thus, we have the 3 possibilities in Figure 1. (We will explain the terms used in Figure 1 soon.)

Figure 1

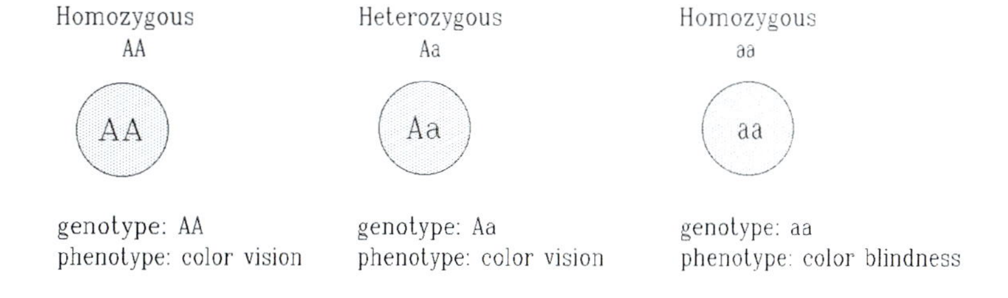

When the alleles in a cell are the same, we say that the cell is **homozygous**, or is a **homozygote**. When the alleles are different, we refer to the cell as **heterozygous**, or as a **heterozygote**.

The types of alleles present in a cell is referred to as the cell's **genotype**. Thus, the cells in Figure 1 have genotypes AA, Aa and aa, respectively. The physical manifestation of the genotype of an organism is referred to as the **phenotype**. (In this case, the phenotype is either color vision or color blindness.)

In the case of the color vision gene, the allele A is **dominant**, which means that if it is present in the body cells, then color vision will be present. On the other hand, the allele a is **recessive**, which implies that a person is color blind if and only if *both* alleles are a, that is, if the person's cells are homozygous with genotype aa.

When organisms reproduce, one allele from each parent cell is duplicated in the child cell, which gives rise to 4 possibilities, as shown in Figure 2. Since each allele is equally likely to be passed on from a given parent to the child, the probability that the child receives any particular pair of alleles is $\frac{1}{4}$.

Figure 2 Each of the following possibilities is equally likely.

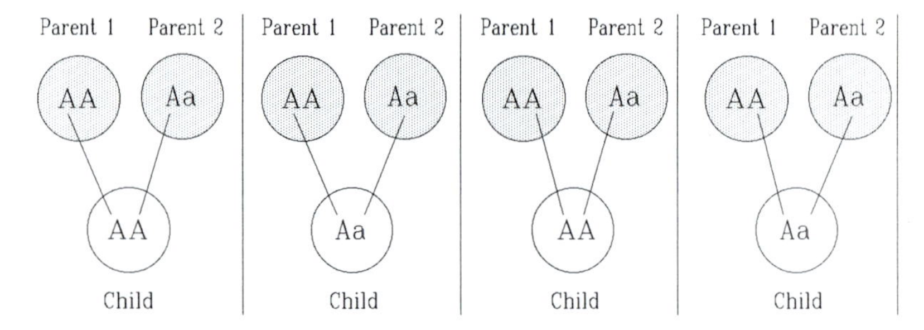

There is often more than one way for a child to receive a certain genotype. For instance, if the parents have genotypes AA and Aa, as in Figure 2, then the child can get genotype Aa in 2 ways, and so the probability that the child is genotype Aa is

$$\frac{1}{4}+\frac{1}{4}=\frac{1}{2}$$

Figure 3 shows the possibilities for single gene matings. To explain the notation, let us consider part d of this figure. In this case, each parent is heterozygous, of genotype Aa. There are, as in all cases, 4 possibilities for the genotype of the offspring, depending on which alleles are passed from the parents. These 4 possibilities are shown in the first part of the figure. The portion after the vertical double bars is a convenient way to summarize the genotype probabilities, and we will use this in the upcoming examples. For instance, the probability that an offspring has genotype Aa is $\frac{1}{2}$, and we denote this by $Aa : \frac{1}{2}$.

Figure 3

a) Parent 1: homozygous AA (color vision), Parent 2: homozygous AA (color vision)

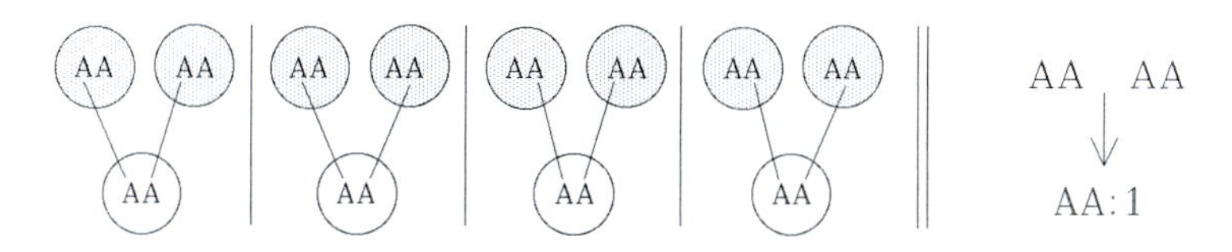

genotype probabilities: $\mathcal{P}(AA) = 1$
phenotype probabilities: $\mathcal{P}(\text{color vision}) = 1$

b) Parent 1: homozygous AA (color vision), Parent 2: heterozygous Aa (color vision)

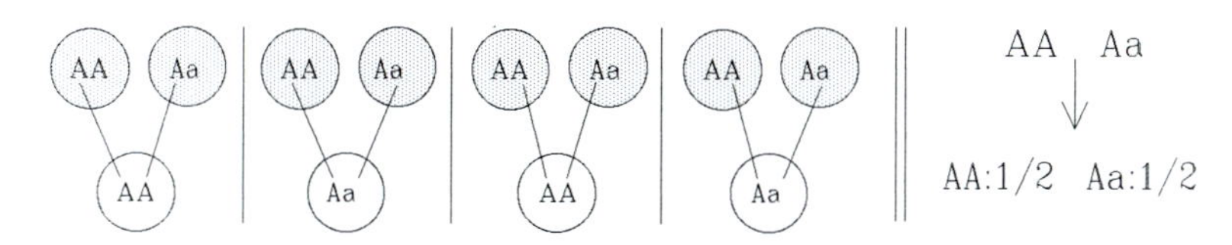

genotype probabilities: $\mathcal{P}(AA) = 1/2, \mathcal{P}(Aa) = 1/2$
phenotype probabilities: $\mathcal{P}(\text{color vision}) = 1$

c) Parent 1: homozygous AA (color vision), Parent 2: homozygous aa (color blindness)

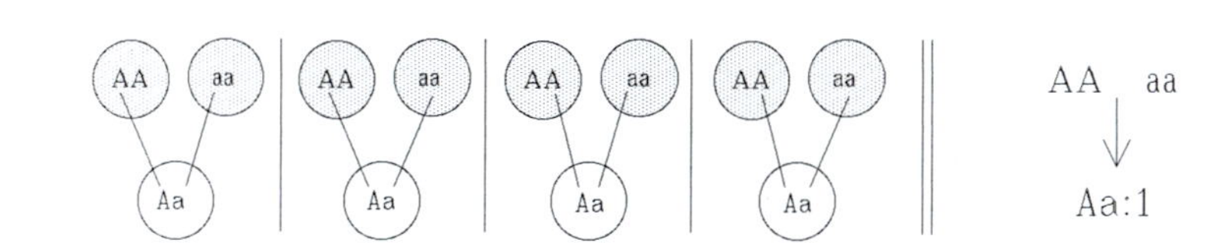

genotype probabilities: $\mathcal{P}(\text{Aa}) = 1$
phenotype probabilities: $\mathcal{P}(\text{color vision}) = 1$

d) Parent 1: heterozygous Aa (color vision), Parent 2: heterozygous Aa (color vision)

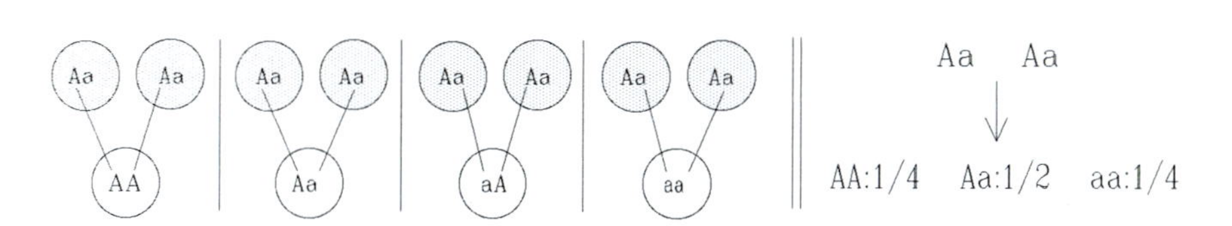

genotype probabilities: $\mathcal{P}(AA) = 1/4, \mathcal{P}(Aa) = 1/2, \mathcal{P}(aa) = 1/4$

phenotype probabilities: $\mathcal{P}(\text{color vision}) = 3/4$, $\mathcal{P}(\text{color blindness}) = 1/4$

e) Parent 1: heterozygous Aa (color vision), Parent 2: homozygous aa (color blindness)

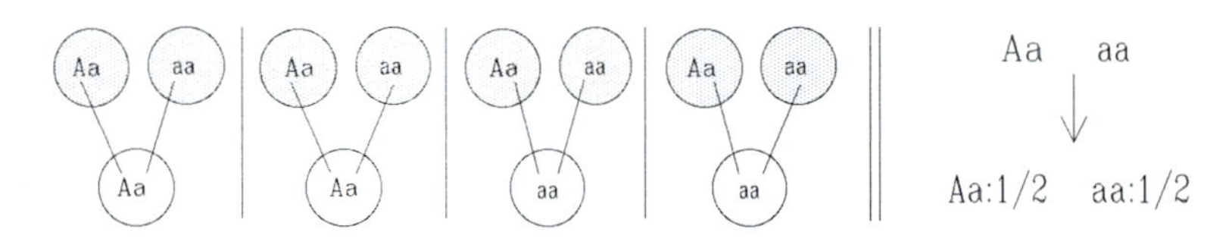

genotype probabilities: $\mathcal{P}(Aa) = 1/2, \mathcal{P}(aa) = 1/2$
phenotype probabilities: $\mathcal{P}(\text{color vision}) = 1/2$, $\mathcal{P}(\text{color blindness}) = 1/2$

f) Parent 1: homozygous aa (color blindness), Parent 2: homozygous aa (color blindness)

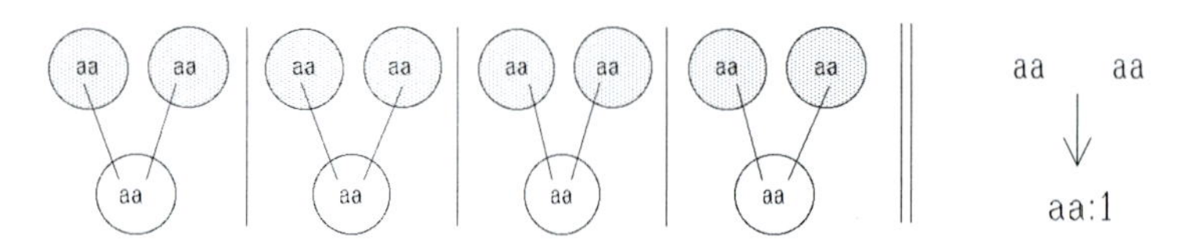

genotype probabilities: $\mathcal{P}(aa) = 1$
phenotype probabilities: $\mathcal{P}(\text{color blindness}) = 1$

Example 1

In a certain line of cats, black hair color is dominant, and white hair color is recessive. Let us denote the dominant allele for black hair by B, and the recessive allele for white hair by b. Suppose that a pure (homozygous) black cat is bred with a white cat, and the resulting litter is allowed to interbreed at random.

a) What is the probability that a given second generation cat will be black?
b) What is the probability that a randomly chosen second generation black cat will be pure black?

Solutions

a) A pure black cat has genotype BB, and a white cat has genotype bb. Hence, their offspring will all have genotype Bb, as shown below

$$\begin{array}{c} \text{BB bb} \\ \downarrow \\ \text{Bb:1} \end{array}$$

If we mate two of these first generation offspring, the resulting genotype probabilities are

$$\begin{array}{c} \text{Bb} \quad \text{Bb} \\ \downarrow \\ \text{BB:}\frac{1}{4} \quad \text{Bb:}\frac{1}{2} \quad \text{bb:}\frac{1}{4} \end{array}$$

Now, a cat is black-haired if its genotype is BB or Bb, and so the probability that a second generation cat will be black is

$$\frac{1}{4} + \frac{1}{2} = \frac{3}{4}$$

b) We can see from part a that $\frac{3}{4}$-th of the second generation cats will be black, and that $\frac{1}{4}$-th of the second generation cats will be pure black (genotype BB.) Hence, the fraction of pure black cats from among all *black* cats is

$$\frac{1/4}{3/4} = \frac{1}{3}$$

In other words, the probability that a randomly chosen black cat is pure black is $\frac{1}{3}$.★

Example 2

A pure (homozygous) black male cat is mated with a heterozygous female cat. A first generation offspring is picked at random, and is mated with a heterozygous cat. What are the genotype probabilities for the resulting second generation offspring?

Solution

A pure black cat has genotype BB, and a heterozygous cat has genotype Bb. Hence, the result of a mating will be

$$\begin{array}{c} \text{BB} \quad \text{Bb} \\ \downarrow \\ \text{BB:}\frac{1}{2} \quad \text{Bb:}\frac{1}{2} \end{array}$$

If we mate a randomly chosen offspring with a heterozygous cat, of genotype Bb, there are two possibilities. If the first generation cat is BB, we have the mating $BB \times Bb$, and if the first generation cat is Bb, we have the mating $Bb \times Bb$. Thus, the results are as follows:

Half the time:	**Half the time:**
BB Bb ↓ BB:$\frac{1}{2}$ Bb:$\frac{1}{2}$	Bb Bb ↓ BB:$\frac{1}{4}$ Bb:$\frac{1}{2}$ bb:$\frac{1}{4}$

From this, we see that an offspring will have genotype BB a total of

$$\frac{1}{2} \cdot \frac{1}{2} + \frac{1}{2} \cdot \frac{1}{4} = \frac{3}{8}$$

of the time. It will have genotype Bb a total of

$$\frac{1}{2} \cdot \frac{1}{2} + \frac{1}{2} \cdot \frac{1}{2} = \frac{1}{2}$$

of the time, and genotype bb a total of

$$\frac{1}{2} \cdot \frac{1}{4} = \frac{1}{8}$$

of the time. In other words, the second generation genotypes are

$$\mathcal{P}(BB) = \frac{3}{8}, \mathcal{P}(Bb) = \frac{1}{2}, \mathcal{P}(bb) = \frac{1}{8} \qquad \bigstar$$

Since producing offspring has no effect on the genotype of the parents (unlike drawing a card from a deck of cards), we see that the genotypes of various offspring of the same parents are independent, and so we may use the definition of independence to help compute probabilities, as the next example illustrates.

Example 3

Two heterozygous black cats (genotype Bb) are mated with each other.

a) What is the probability that, if the litter has size 3, the kittens will be born in the order white-black-black?

b) What is the probability that, if the litter has size 3, it will contain 2 black kittens and 1 white kitten?

Solutions

a) Referring to Figure 3d, we know that the probabilities of a black and a white kitten are

$$\mathcal{P}(\text{black}) = \frac{3}{4}, \mathcal{P}(\text{white}) = \frac{1}{4}$$

Hence, since the genotype of the three offspring are independent, the probability of a white-black-black litter is

$$\mathcal{P}(\text{white-black-black}) = \frac{1}{4} \cdot \frac{3}{4} \cdot \frac{3}{4} = \frac{9}{64}$$

b) If we think of the outcome of getting a black kitten as a success, then we have a binomial experiment, and we want the probability of exactly 2 successes. Theorem 2 of Section 2.2 tells us that this is

$$\mathcal{P}(\text{2 black, 1 white}) = C(3,2)\left(\frac{3}{4}\right)^2\left(\frac{1}{4}\right) = \frac{27}{64} \qquad \bigstar$$

Testcrossing

If an allele a is recessive to a dominant allele A, then there is no way to tell the genotypes AA and Aa simply by looking at the phenotype of the organism, since its appearance is determined by the presence of the dominant allele. However, we can get some information about the genotype by a procedure known as *testcrossing*. This is simply the process of mating the organism in question with a homozygous recessive organism (genotype aa). If any one of the resulting offspring have the recessive phenotype, we know that the offspring has genotype aa, and so each of the parents must have at least one recessive allele. This tells us that the parent in question must be heterozygous (genotype Aa). On the other hand, if all of the offspring have the dominant phenotype, then we can only make some probabilistic statements about the genotype of the parent in question. Let us illustrate.

Example 4

Suppose we breed a black cat with a white cat, and produce a litter of 5 black cats. What can we say about the genotype of the black parent?

Solution

We know that the genotype of the black parent must be either BB or Bb, since it has phenotype black. Also, the genotype of the white cat is bb. Let us compute the probability that, if the black parent has genotype Bb, it would produce a litter of 5 black cats.

We know from Figure 3 that the probability of a single black offspring from a mating pair of the form $Bb \times bb$ is $\frac{1}{2}$. Hence, the probability of 5 black offspring is

$$\frac{1}{2} \times \frac{1}{2} \times \frac{1}{2} \times \frac{1}{2} \times \frac{1}{2} = \frac{1}{32} \approx 0.031 \qquad (1)$$

Thus, if the black parent was heterozygous Bb, the chances of a 5-kitten all black litter is quite remote, and so we can be fairly confident that the parent is homozygous. However, it is very important to emphasize that we cannot be certain that the parent is homozygous, for the next kitten may indeed be white. The point is that, the smaller the number computed in (1) is, the more confident we can be of a homozygous parent. ★

EXERCISES

Define or discuss each of the following terms.

a) Gene
b) Allele
c) Zygote
d) Homozygote
e) Heterozygote
f) Genotype
g) Phenotype
h) Dominant allele
i) Recessive allele
j) Independent events
k) Testcrossing

1. In the fruit fly, brown eyes are determined by a recessive gene a, whereas red eyes are determined by a dominant gene A. Suppose that a pure red-eyed fruit fly is crossed with a brown-eyed fruit fly, and the first generation is allowed to interbreed at random.
   a) What is the probability that a randomly chosen second generation fruit fly will be red-eyed?
   b) What is the probability that a randomly chosen second generation fruit fly will be brown-eyed?
   c) What is the probability that a randomly chosen second generation red-eyed fruit fly will be homozygous?
2. In rabbits, short hair is determined by a dominant allele S, whereas long hair is determined by a recessive allele s. Suppose that a pure short-haired rabbit is bred with a long-haired rabbit, and an offspring is picked at random, and mated with a long-haired rabbit.
   a) What is the probability that a randomly chosen second generation rabbit will be short-haired?
   b) What is the probability that a randomly chosen second generation rabbit will be long-haired?
   c) What is the probability that a randomly chosen second generation short-haired rabbit will be homozygous?
3. In a certain species of fox, a silver coat c is recessive, and a red coat C is dominant. Suppose a pure red fox is mated with a heterozygous fox, and then an offspring is selected at random and mated with a silver fox. What are the genotype probabilities of the second generation?
4. In a certain species of fox, a silver coat c is recessive, and a red coat C is dominant. Suppose a pure red fox is mated with a heterozygous fox, and then an offspring is selected at random and mated with a pure red fox. What are the genotype probabilities of the second generation?
5. Red hair in humans is recessive, whereas dark hair is dominant. Suppose that a red-haired male and a heterozygous female have a male child who then mates with a heterozygous female.
   a) What is the probability that the grandchild is pure dark-haired?
   b) What is the probability that the grandchild is dark-haired?
   c) What is the probability that the grandchild is red-haired?
6. Bushy eyebrows in humans is recessive, whereas normal eyebrows is dominant. Suppose that a male with bushy eyebrows and a heterozygous female have a female child who then mates with a heterozygous male.
   a) What is the probability that the grandchild has bushy eyebrows?
   b) What is the probability that the grandchild is pure normal eyebrowed?
   c) What is the probability that the grandchild has normal eyebrows?
7. In rabbits, a spotted coat is dominant, whereas a solid coat is recessive. A pure spotted rabbit is bred with a heterozygous rabbit. A first generation offspring is picked at random, and is mated with a solid-coated rabbit. What are the genotype probabilities for the resulting second generation offspring?
8. In sheep, white wool is dominant, whereas black wool is recessive. A pure white sheep is bred with a heterozygous sheep. A first generation offspring is picked at random, and is mated with a heterozygous sheep. What are the genotype probabilities for the resulting second generation offspring?

9. Hemophilia is a disease in humans that manifests itself in the form of excessive bleeding. Fortunately, the allele for hemophilia is recessive. A man with no hemophilia alleles mates with a woman who is heterozygous for hemophilia. Their son mates with a heterozygous woman.
   a) What is the probability that the grandchild has hemophilia?
   b) What is the probability that the grandchild is homozygous normal?
   c) What is the probability that the grandchild is heterozygous for hemophilia?
10. Two heterozygous black cats mate.
    a) What is the probability that, if the litter has size 2, the kittens will be born in the order white-black?
    b) What is the probability that, if the litter has size 2, it will contain 1 black kitten and 1 white kitten?
11. Red hair in humans is recessive, and dark hair is dominant. Two heterozygous humans decide to raise a family of 3 children.
    a) What is the probability that the children will all be red-haired?
    b) What is the probability that none of the children will be red-haired?
    c) What is the probability that exactly 1 of the children will be red-haired?
    d) What is the probability that exactly 2 of the children will be red-haired?
12. Red hair in humans is recessive, and dark hair is dominant. A red-haired man and a heterozygous woman decide to raise a family of 3 children.
    a) What is the probability that the children will all be red-haired?
    b) What is the probability that none of the children will be red-haired?
    c) What is the probability that exactly 1 of the children will be red-haired?
    d) What is the probability that exactly 2 of the children will be red-haired?
13. Short hair in rabbits is a dominant trait. Two heterozygous rabbits are bred, and produce 3 offspring.
    a) What is the probability that the rabbits are born in the order short-long-short?
    b) What is the probability that 2 of the offspring are short-haired, and 1 is long-haired?
    c) What is the probability that all 3 of the offspring are short-haired?
    d) What is the probability that all 3 of the offspring are long-haired?
14. Color blindness in humans is recessive. A heterozygous man and a woman who is homozygous for color blindness decide to raise a family of 3 children.
    a) What is the probability that 2 of the children have normal vision, and 1 is color blind?
    b) What is the probability that 1 of the children has normal vision, and 2 are color blind?
    c) What is the probability that all of the children have normal vision?
    d) What is the probability that all of the children are color blind?

15. Wire-like hair is a dominant trait in a certain species of dogs, whereas smooth hair is recessive. Two heterozygous dogs are bred, and produce a litter of size 4.
    a) What is the probability that the puppies are born in the order wire-wire-smooth-smooth?
    b) What is the probability that 3 of the puppies are wire-haired, and 1 is smooth-haired?
    c) What is the probability that exactly 2 of the puppies are wire-haired?
    d) What is the probability that all 4 of the puppies are smooth-haired?
16. Suppose we breed a black cat with a white cat, and produce a litter of 4 black cats. What can we say about the genotype of the black parent, and how confident do we feel about this conclusion?
17. Suppose that a red-haired man and a black-haired woman have 2 children, both of whom are black-haired. (Red hair is recessive, black hair is dominant.) Can we draw any strong conclusions about the genotype of the woman?

Codominance

In some traits, alleles are neither dominant nor recessive, but always express themselves to some extent in the phenotype. Such alleles are referred to as **codominant.** For codominant alleles, we use a notation such as A^1, A^2.

18. The M-N blood group in humans is codominant. The customary notation for these alleles is L^M and L^N, where the L is in honor of the discoverers of this blood group (Karl Landsteiner and Philip Levine.) The possible phenotypes are M (genotype L^ML^M), N (genotype L^NL^N), and MN (genotype L^ML^N). Suppose that a couple, both of whom are blood group MN, have a child, who mates with a person of blood group M. What are the probabilities of the various phenotypes for the grandchild?
19. Referring to the previous exercise, suppose that a couple, both of whom are blood group MN, have a child, who mates with a person of blood group MN. What are the probabilities of the various phenotypes for the grandchild?

Answers To Odd Numbered Exercises

Section 1.1

3. $6 \cdot 4 \cdot 3 \cdot 10 = 720$ 5. $3 \cdot 4 \cdot 6 \cdot 3 = 216$
7. 15 9. $4 \cdot 3 \cdot 2 = 24$
11. $8,100,000,000$ 13. $4^{10} = 1,048,576$
15. 12
17. a) $78^4 = 37,015,056$ b) $78 \cdot 77 \cdot 76 \cdot 75 = 34,234,200$
19. a) 11 b) 40
21. a) 240 b) 840
23. $5 \cdot 6 \cdot 6 \cdot 3 = 540$

Section 1.2

1. a)1 b)1 c)6 d)24 e)120 f)720
3. a) 95040 b) $5,079,110,400$ c) 94,109,400
5. a)3 b)4 c)5 d)n
7. no; take $n = 1$, $m = 2$
9. 13!; world population $\approx 5,250,000,000$;
   $13! = 6,227,020,800$, $12! = 479,001,600$
11. 2; $12, 21$
13. $P(6,5) = 6 \cdot 5 \cdot 4 \cdot 3 \cdot 2 = 720$
15. $P(5,2) = 20; 12, 13, 14, 15, 21, 23, 24, 25, 31, 32, 34, 35,$
    $41, 42, 43, 45, 51, 52, 53, 54$
17. $10! = 3,628,800$; approximately 6.9 years
19. $5! \cdot 20! \approx 2.92 \times 10^{20}$
21. $P(9,3) = 9 \cdot 8 \cdot 7 = 504$
23. $7! = 5040$
25. $2 \cdot 4! \cdot 4! = 1152$
27. $P(25,5) + 4 \cdot P(24,3) = 6,424,176$
29. $26^3 = 17,576$
31. a) $2 \cdot 26^3 = 35,152$ b) $2 \cdot 25 \cdot 24 \cdot 23 = 27,600$
    c) $2 \cdot 26 \cdot 25 \cdot 24 = 31,200$
33. $5! \cdot 5! \cdot 5! = 1,728,000$
35. $P(10,5) = 30,240$
37. a) $2^8 = 256$ b) 2 c) 6

Section 1.3

1. 2 3. 3 5. 5 7. 6 9. 20 11. 56 13. 36 15. 120
17. 210 19. 3003 21. 15 23. 77520 25. 201,359,550
27. 75 29. 4950
35. $10; \{1,2,3\}, \{1,2,4\}, \{1,2,5\}, \{1,3,4\},$
    $\{1,3,5\}, \{1,4,5\}, \{2,3,4\}, \{2,3,5\}, \{2,4,5\}, \{3,4,5\}$
37. $C(55,5) = 3,478,761$
39. $C(10,5) \cdot C(10,5) = 63,504; C(10,5) = 252$
41. a) $2^4 = 16$ b) $C(4,1) = 4$
43. a) $2^7 = 128$ b) $C(7,3) = 35$
45. a) $3^4 = 81$ b) $C(4,2) \cdot 2 \cdot 2 = 24$

47. $C(9,4) + C(8,4) = 196$
49. a) $C(9,6) + C(9,5) + C(9,5) = 336$ b) $2 \cdot C(6,3) = 40$
51. $n = 15$, since $C(14,3) = 364 < 365$, but $C(15,3) = 455 > 365$

Section 2.1

1. a) $\frac{1}{13}$ b) $\frac{1}{4}$ c) $\frac{1}{2}$
3. a) $\frac{1}{6}$ b) $\frac{1}{2}$ c) $\frac{1}{2}$
5. $\mathcal{P}(\text{heads}) = \frac{1}{4}$, $\mathcal{P}(\text{tails}) = \frac{3}{4}$
7. $\mathcal{P}(1), \ldots, \mathcal{P}(7)$ equal $\frac{1}{13}$, $\mathcal{P}(8) = \mathcal{P}(9) = \mathcal{P}(10) = \frac{2}{13}$; $\mathcal{P}(\text{even}) = \frac{7}{13}$
9. $\frac{1}{2}$ 11. $\frac{3}{8}$
13. a) $\frac{5}{12}$ b) $\frac{9}{12}$ c) $\frac{7}{12}$
15. Sample space is $S = \{\text{red, blue, black}\}$, $\mathcal{P}(\text{red}) = \frac{5}{21}$, $\mathcal{P}(\text{blue}) = \frac{6}{21}$, $\mathcal{P}(\text{black}) = \frac{10}{21}$, $\mathcal{P}(\text{not red}) = \frac{16}{21}$.
17. Area of inner region $= \pi$. Area of middle region $= 4\pi - \pi = 3\pi$. Area of the outer region is $9\pi - 4\pi = 5\pi$. Area of entire board is 9π. $\mathcal{P}(\text{hitting inner region}) = \pi/9\pi = 1/9$. $\mathcal{P}(\text{hitting middle region}) = 3\pi/9\pi = 1/3$. $\mathcal{P}(\text{hitting outer region}) = 5\pi/9\pi = 5/9$.
19. $\frac{48}{C(52,5)} \approx 0.00002$
21. $\frac{C(13,1) \cdot 48}{C(52,5)} \approx 0.00024$
23. $\frac{C(13,5)}{C(52,5)} \approx 0.000495$
25. Ordered pairs whose components are from the set $\{1,2,3,4\}$; 16
27. $\frac{3}{4}$ 29. $\frac{1}{2}$ 31. $\frac{24}{6 \cdot 52} = \frac{1}{13}$

Section 2.2

1. $1170/2652 \approx 0.4412$
3. $396/2652 \approx 0.1493$
5. $1023/1024 \approx 0.9990$
7. $5/6$
9. This is the event of getting a diamond or a card with face value at least 9; $\{A\diamondsuit, 2\diamondsuit, \ldots, K\diamondsuit, 9\clubsuit, 10\clubsuit, J\clubsuit, Q\clubsuit, K\clubsuit, 9\heartsuit, 10\heartsuit, J\heartsuit, Q\heartsuit, K\heartsuit, 9\spadesuit, 10\spadesuit, J\spadesuit, Q\spadesuit, K\spadesuit\}$
11. This is the event of getting a face card or a card of face value at least 9. But this is the same as the event H of getting a card with face value at least 9. That is, $F \cup H = H =$ $\{9\clubsuit, 10\clubsuit, J\clubsuit, Q\clubsuit, K\clubsuit, 9\diamondsuit, 10\diamondsuit, J\diamondsuit, Q\diamondsuit, K\diamondsuit, 9\heartsuit, 10\heartsuit, J\heartsuit, Q\heartsuit, K\heartsuit, 9\spadesuit, 10\spadesuit, J\spadesuit, Q\spadesuit, K\spadesuit\}$
13. This is the event of getting a diamond or a face card, provided that it has face value at least 9; $\{9\diamondsuit, 10\diamondsuit, J\diamondsuit, Q\diamondsuit, K\diamondsuit, J\clubsuit, Q\clubsuit, K\clubsuit, J\heartsuit, Q\heartsuit, K\heartsuit, J\spadesuit, Q\spadesuit, K\spadesuit\}$
15. This is the event of getting a sum of 7 and an even sum. But this is the empty event $\emptyset$, since E and F are mutually exclusive.
17. This is the event of getting an even sum, with at least one die odd. But for the sum of two numbers to be even when one number is odd, the other number must also be odd, and so this event is the set of all ordered pairs

where both numbers are odd;

$$\{(1,1),(1,3),(1,5),(3,1),(3,3),(3,5),(5,1),(5,3),(5,5)\}$$

19. This is the event of getting a sum of 7 or an outcome where at least one die is odd. But, getting a sum of 7 can only happen if one of the numbers is odd (and the other is even), and so $E \subset H$. Hence, $E \cup H = H$.
21. $11/26$ 23. a) $1/4$ b) $3/4$ 25. a) $1/8$ b) $5/8$
27. $5/18$ 29. Independent 31. Independent
33. Yes, regardless of whether or not you get a heads, the probability of getting an ace is $\frac{2}{26}$.
35. $\mathcal{P}(3 \text{ on die, tails on coin})$
$= \mathcal{P}(3 \text{ on die}) \cdot \mathcal{P}(\text{tails on coin}) = \frac{1}{6} \cdot \frac{1}{2} = \frac{1}{12}$
37. a) $\frac{1}{52^4} \approx 0.000000136$ b) $\frac{1}{13} \cdot \frac{1}{13} \cdot \frac{1}{13} \cdot \frac{1}{13} \approx 0.000035$ 39. $19/100$
41. $\frac{C(4,1)}{2^4} = \frac{1}{4}$ 43. $\frac{C(4,3)}{2^4} = \frac{1}{4}$ 45. $\frac{C(4,3)}{2^4} + \frac{C(4,4)}{2^4} = \frac{5}{16}$
47. $\mathcal{P}(\text{at least 1 head}) = 1 - \mathcal{P}(\text{no heads}) = 1 - \frac{C(4,0)}{2^4} = \frac{15}{16}$
49. $C(5,1)\left(\frac{1}{6}\right)^1\left(\frac{5}{6}\right)^4 \approx 0.4019$
51. $C(5,3)\left(\frac{1}{6}\right)^3\left(\frac{5}{6}\right)^2 \approx 0.0322$
53. $C(5,5)\left(\frac{1}{6}\right)^5\left(\frac{5}{6}\right)^0 \approx 0.0001$
55. $C(5,3)\left(\frac{1}{6}\right)^3\left(\frac{5}{6}\right)^2 + C(5,4)\left(\frac{1}{6}\right)^4\left(\frac{5}{6}\right)^1 + C(5,5)\left(\frac{1}{6}\right)^5\left(\frac{5}{6}\right)^0 \approx 0.0355$
57. $C(4,1)\left(\frac{1}{13}\right)^1\left(\frac{12}{13}\right)^3 \approx 0.2420$
59. $C(4,2)\left(\frac{1}{13}\right)^2\left(\frac{12}{13}\right)^2 + C(4,3)\left(\frac{1}{13}\right)^3\left(\frac{12}{13}\right)^1 + C(4,4)\left(\frac{1}{13}\right)^4\left(\frac{12}{13}\right)^0 \approx 0.03197$
61. $C(4,2)\left(\frac{1}{4}\right)^2\left(\frac{3}{4}\right)^2 = \frac{27}{128}$
63. $C(4,3)\left(\frac{1}{4}\right)^3\left(\frac{3}{4}\right)^1 + C(4,4)\left(\frac{1}{4}\right)^4\left(\frac{3}{4}\right)^0 = \frac{13}{256}$
65. $2/9$
67. a) No, since the drawing of socks is not independent.
b) $\frac{4 \cdot 6 \cdot 3!}{8 \cdot 7 \cdot 6} \approx 0.4286$
69. 0 (he cannot end up in front of the hospital)
71. $\frac{C(n,k)}{2^n}$

Section 2.3

1. $\frac{1}{6}$ 3. $\frac{1}{3}$ 5. $\frac{1}{6}$ 7. $\frac{1}{9}$ 9. $\frac{5}{18}$ 11. $\frac{1}{6}$ 13. $\frac{1}{4}$ 15. $\frac{1}{4}$
17. $\frac{4}{51}$ 19. $\frac{1}{13}$ 21. $\frac{1}{33}$ 23. $\frac{1}{4}$ 25. $\frac{1}{13}$ 27. 0 29. $\frac{1}{4}$
31. $\frac{1}{4}$ 33. (a) 0.014 (b) 0.0525 (c) 0.8455
35. $\mathcal{P}(\text{green}) = \frac{11}{24}$, $\mathcal{P}(\text{yellow}) = \frac{13}{24}$ 37. $\frac{4}{9}$ 39. 0.000325
41. $\frac{1}{2}$ dollar; yes, the game is worth playing since the expected value is positive
43. $\frac{25}{13}$ cents ; yes, the game is worth playing since the expected value is positive
45. 0.5 cents; yes, the game is worth playing since the expected value is positive
47. 8.5 cents 49. $\mathcal{E} \approx 5.56$ cents; yes 51. yes; $\mathcal{E} = 0$

Section 3.1

1. $\frac{4}{C(52,5)} \approx 0.000001539$
3. $\frac{4 \cdot C(13,5) - 40}{C(52,5)} = 0.001965401$
(we subtract 40 since there are 40 straight and royal flushes)

5. $\frac{13 \cdot C(4,2) \cdot C(12,3) \cdot 4^3}{C(52,5)} \approx 0.422569027$
7. nothing
9. $\frac{4}{47}$ 11. $\frac{1}{47}$ 13. $\frac{46}{C(47,2)} = \frac{2}{47}$ 15. $\frac{15}{47}$
17. $\frac{C(6,3) \cdot C(47,3)}{C(53,6)} \approx 0.014126114$
19. $\frac{C(6,1) \cdot C(47,5)}{C(53,6)} \approx 0.40089914$
21. $\mathcal{P}(\text{losing}) \approx 0.985155255$
23. -0.3595103 dollars
25. $-\ 0.491633042$ dollars

Section 3.2

1. a) $\frac{3}{4}$ b) $\frac{1}{4}$ c) $\frac{1}{3}$ 3. $Cc : \frac{3}{4}, cc : \frac{1}{4}$
5. a) $\frac{1}{8}$ b) $\frac{5}{8}$ c) $\frac{3}{8}$ 7. $Ss : \frac{3}{4}, ss : \frac{1}{4}$
9. a) $\frac{1}{8}$ b) $\frac{3}{8}$ c) $\frac{1}{2}$
11. a) $\frac{1}{4^3} = \frac{1}{64}$ b) $\left(\frac{3}{4}\right)^3 = \frac{27}{64}$ c) $3 \cdot \frac{1}{4} \cdot \left(\frac{3}{4}\right)^2 = \frac{27}{64}$
    d) $3 \cdot \left(\frac{1}{4}\right)^2 \left(\frac{3}{4}\right) = \frac{9}{64}$
13. a) $\frac{3}{4} \cdot \frac{1}{4} \cdot \frac{3}{4} = \frac{9}{64}$ b) $C(3,1) \cdot \frac{1}{4} \left(\frac{3}{4}\right)^2 = \frac{27}{64}$ c) $\left(\frac{3}{4}\right)^3 = \frac{27}{64}$
    d) $\left(\frac{1}{4}\right)^3 = \frac{1}{64}$
15. a) $\frac{3}{4} \cdot \frac{3}{4} \cdot \frac{1}{4} \cdot \frac{1}{4} = \frac{9}{256}$ b) $C(4,1) \cdot \frac{1}{4} \cdot \left(\frac{3}{4}\right)^3 = \frac{27}{64}$
    c) $C(4,2) \left(\frac{3}{4}\right)^2 \left(\frac{1}{4}\right)^2 = \frac{27}{128}$ d) $\left(\frac{1}{4}\right)^4 = \frac{1}{256}$
17. No
19. $\mathcal{P}(\text{group M}) = \frac{1}{4}$, $\mathcal{P}(\text{group MN}) = \frac{1}{2}$, $\mathcal{P}(\text{group N}) = \frac{1}{4}$